GOAL

QUALITY OF LIFE
PRODUCTION AND LANDSCAPE DESCRIPTION

ECOSYSTEM FOUNDATION BLOCKS

Water Cycle	Mineral Cycle	Energy Flow

Succession

TOOLS

Rest	Fire	Grazing	Animal Impact	Living Organisms	Technology	Money & Labor

Human Creativity

GUIDELINES

Whole Eco-System	Weak Link	Cause & Effect	Marginal Reaction	Energy/ Wealth Source & Use	Society & Culture	Time	Stock Density	Herd Effect	Population Management	Burning	Flexibility -Strategic -Tactical -Operational	Biological Plan Monitor Control Replan	Organization/ Personal Growth	$ Plan Monitor Control Replan

Gross Margin Analysis

TESTING

MANAGEMENT

Holistic resource management model.

HOLISTIC RESOURCE MANAGEMENT
Workbook

Sam Bingham with Allan Savory

Island Press
Washington, D.C. • Covelo, California

ABOUT ISLAND PRESS

Island Press, a nonprofit organization, publishes, markets, and distributes the most advanced thinking on the conservation of our natural resources—books about soil, land, water, forests, wildlife, and hazardous and toxic wastes. These books are practical tools used by public officials, business and industry leaders, natural resource managers, and concerned citizens working to solve both local and global resource problems.

Founded in 1978, Island Press reorganized in 1984 to meet the increasing demand for substantive books on all resource-related issues. Island Press publishes and distributes under its own imprint and offers these services to other nonprofit organizations.

Support for Island Press is provided by Apple Computers, Inc., Mary Reynolds Babcock Foundation, Geraldine R. Dodge Foundation, The Educational Foundation of America, The Charles Engelhard Foundation, The Ford Foundation, Glen Eagles Foundation, The George Gund Foundation, The William and Flora Hewlett Foundation, The Joyce Foundation, The J.M. Kaplan Fund, and John D. and Catherine T. MacArthur Foundation, The Andrew W. Mellon Foundation, The Joyce Mertz-Gilmore Foundation, The New-Land Foundation, The Jessie Smith Noyes Foundation, The J.N. Pew, Jr., Charitable Trust, Alida Rockefeller, The Rockefeller Brothers Fund, The Florence and John Schumann Foundation, The Tides Foundation, and individual donors.

For additional information about Island Press publishing services and a catalog of current and forthcoming titles, contact Island Press, Box 7, Covelo, California 95428: 1-800-828-1302.

Design and Production: TypArt, Ltd, Albuquerque, New Mexico

Printed on recycled, acid-free paper

Library of Congress Cataloging-in-Publication Data

Bingham, Sam
  The holistic resource management workbook / Sam Bingham with Allan Savory.
    p.   cm.
  ISBN 0-933280-69-6 (alk. paper)
  1. Range management.   2. Grazing—Management.   3. Land Use—Planning.
  I. Savory, Allan, 1935- . II. Title.
  SF85.B48 1990
  333.74—dc20   89-24488
                                                         CIP

MANUFACTURED IN THE UNITED STATES OF AMERICA

10 9 8 7 6 5 4 3 2 1

CONTENTS

Foreword iv
Preface v
Acknowledgments ix

Part I: FINANCIAL PLANNING
FINANCIAL PLANNING 3
MASTERING THE BASICS 4
  Goals 4
  Guidelines 4
  Good Ideas 15
CREATING YOUR PLAN 21
  Planning the Planning 21
  Planning the Income 23
  Planning the Expenses 25
  Managing the Cash 30
  Analyzing the Plan 32
  Control Through the Year 32
SUMMARY 35

Part II: BIOLOGICAL PLANNING
BIOLOGICAL PLANNING 39
MASTERING THE BASICS 40
  Measuring Range Use and Forage 40
  Grazing and Growth Rates 44
  Grazing Periods and Recovery Periods 45
  Time, Paddocks, and Land Divisions 45
  Forage Reserves 50
  A Final Word About Stocking Rate 53
  The Critical Dormant Season 55
  Herd Effect 61
  Multiple Herds 62
  Matching Animal Cycles to Land Cycles 63
  Pests, Parasites, and Other Headaches 66
CREATING YOUR PLAN 67
  The *Aide Memoire* for Biological Planning 68
  Guidelines for Simplified Grazing Management 80
SUMMARY 83

Part III: BIOLOGICAL MONITORING
BIOLOGICAL MONITORING 87
MASTERING THE BASICS 89
  General Observations 89
  Succession 90
  Water Cycle 92
  Mineral Cycle 94
  Energy Flow 96
  Plant Forms 98
  Soil Capping 100
  Identifying Species 100
  Grazing Patterns 102
  Living Organisms 104
CREATING YOUR PLAN 106
  Gathering the Data 106
  Summarizing the Data 115
SUMMARY 120

Part IV: LAND PLANNING
LAND PLANNING 123
MASTERING THE BASICS 125
  Collecting Information 125
  Preparing Maps and Overlays 126
  Deciding on Herd and Cell Sizes 132
CREATING YOUR PLAN 136
  Figuring Costs and Schedules 139
  Layouts and Hardware 146
EPILOGUE 157
Appendices
  A. Financial Planning Forms 159
  B. Biological Planning Forms 164
  C. Biological Monitoring Forms 166
  D. The Center for Holistic Resource Management 171
Glossary 172
Index 175

FOREWORD

Four years ago when outlining chapters for a textbook on holistic resource management, I realized it would be more effective in two parts: textbook and workbook. The general theory and the emerging possibilities for holistic management needed to be understood by a general readership. Not only do these new ideas touch all of our lives profoundly, they provide ammunition to effect the changes needed to ensure our planet's survival. Thus I wrote *Holistic Resource Management* as a standard text for a broad readership.

On the other hand, the practical application of holistic principles on the land involves a lot of technical detail that could cloud the understanding of readers interested primarily in the theoretical aspects of holistic management. I had reservations about writing this workbook myself, mainly because I had grown stale in presenting material I had worked with for thirty years and knew like the back of my hand. A fresh approach was needed—authored by someone new enough to the subject to appreciate the difficulties encountered by beginners. I found such a person in Sam Bingham. He understood the struggles of those new to practicing holistic management because he had experienced them himself over a six-year period. And as a writer he had a flair, which I lacked, for making somewhat dry material come to life.

When I first met Sam eight years ago, he was a determined young man teaching Navajo children about resource management in a way that combined current science with the treasure house of knowledge in their own culture. Struck by his enthusiasm and his openness, I invited him to attend a course I was teaching a few months later. Using what he learned at that course and the methods he developed to teach the grazing aspects of holistic management to his Navajo students, he wrote a marvelous book called *Living from Livestock*.

I was so impressed by its simplicity, and the clarity of the illustrations he had devised, that I determined to enlist his talents in the writing of my own book. This I fortunately was able to do, thanks to funding provided by the 777 Fund of the Tides Foundation. Sam helped in the writing and editing of the HRM textbook and is the principal author of this workbook.

Sam had a free hand in the organization, design, and illustration. My only instructions were to keep it simple. As you will see, he succeeded far better than I dared hope. Sam has produced a book that is not only clear, readable, and practical but adventurous, funny, and wise. I hope you'll find this book as enlightening and useful as I have. As always, we invite your comments, criticism, and suggestions for improvement. Your contributions and practical experience have brought us this far and remain essential to advancing the frontiers of holistic management.

Allan Savory

PREFACE

This book is designed as a practical companion for people who actually manage land and any others who want to know what responsibility for land really means. Most of all, however, it is dedicated to the farmers, ranchers, tribes, and villagers who make a living directly from the earth itself. Upon them rests the fate of everyone else, including myself, a writer who despite a long interest in agriculture has only in extreme cases had to eat what he produced.

Allan Savory, however, whose ideas this book represents, nevertheless insists that no mind stands more open to learning than an empty one—and indeed I have tried to make this book not mine at all but the work of many practical people whose combined experience encompasses centuries of time on the land. My own experience with the ideas that have since become known as holistic resource management goes back to 1980 when my wife and I were teaching on the Navajo Indian Reservation in Arizona, and the local school board commissioned us to write a history of their country. They told us, however, not to organize it around the warriors and politicians that usually populate history texts but around the changing relationship between people and their land.

In our first interview we asked one of the oldest inhabitants to cook us the food she ate as a child. She smiled at our innocence.

"Oh, I couldn't do that," she said. "The plants we ate just don't exist anymore. When the government reduced our livestock, the sky was insulted and cut off the rain. The plant people were insulted and simply left. We'd starve, too, if the government didn't print food stamps. Modern children grow up eating paper."

She gestured toward a valley, which but for the occasional saltbush had the surface of a good tennis court, and said, "I used to go up there in the morning and by noon gather enough grass seed to feed my family for a day. You try that now, young man."

As teacher of the high school agriculture course, I felt I had to have a scientific explanation, but higher-level experts offered nothing I could easily adapt for my Navajo students. The government range conservationists all shrugged and said, "The Navajos cheat on their sheep permits. What can you expect?" Most added that our friend's high grass never grew there anyway except in a senile imagination. None offered any hope of restoring it.

About that time an anthropologist named Rosalie Finale joined a U.S. Park Service research project at Chaco Canyon National Monument in the middle of Navajo Country. She had recently participated in a multi-disciplinary study of a catastrophic drought in the Sahel region of West Africa and wanted to prove yet another theory.

She believed that social disruption destroyed the Navajo range. It had little to do with animal numbers or the sky's reluctance to rain, she said, but came from destroying traditional grazing patterns through the establishment of grazing districts, permits, dispersed water development, restrictions on moving herds, and other regulations that attended stock reduction. The social and biological scientists on her team in Africa had divided sharply over the suggestion that politics could dry up grass in such vast quantities. Indeed, she could not explain the biological principles involved, but she made a good case from circumstantial evidence.

We were fortunate to have no theory of our own, because the old woman and the anthropologist, each in her own way, forced us to suspect that we would only solve the riddle of the Navajo desert through a very unorthodox mixture of fact and theory. Then someone sent us an article about Allan Savory. Here at last was an apparently practical thinker with a broad view.

Without any introduction, we called him on the telephone, invited him to Navajo Country, and he came. The old woman's story made perfect sense to him. Unlike the experts before him, he actually checked it on the ground and found in faint hummocks and bits of root among the moribund saltbushes proof that alkali sacaton grass no doubt once did yield seed in abundance. He confirmed the link between the social disruption of stock reduction and desertification and recast the story about stock reduction and the insulted sky into a scientific metaphor. That eventually made vast areas of the older generations' experience accessible to the school-trained minds of our students—and ourselves.

After Savory's public speech later that day, Navajos old and young applauded him as the first "expert" who made enough sense to offer any hope. With all the zeal of new converts, some Navajo friends and I asked American taxpayers to build us a genuine Savory grazing cell complete with miles of wire and a 5,000-volt fence charger from New Zealand. Eventually they did. They also underwrote a local trial of keeping half-wild Navajo angora goats inside electric fence.

We never did manage to contain the goats, and the cell has yet to perform as it should. Only gradually did we see that we had failed by not mastering either holism or the mental attitude needed to apply it. Meanwhile it offended our pride to have to phone Savory for advice at every turn, even after attending his courses.

As it turned out, dependence such as ours troubled Allan too, and he worked hard to find better ways to articulate his particular way of analyzing what he saw on the ground. His efforts eventually crystallized around the concept of holism, which he came to see as the fourth missing key to understanding the ecosystem.

Later Allan asked me to work with him on his textbook *Holistic Resource Management*, which led to months of intense discussion about how to express this discovery. In the process Allan, his other editor (Jody Butterfield, coincidently also his wife), and I came to realize how extraordinarily difficult it is, using mere words, to hack a pathway through the thicket of custom and prior lessons that keeps a good theory from enlightening practice.

The rest of *Holistic Resource Management* was almost ready for production before we hammered out a comprehensible draft of the key chapter on holism. We rewrote the first attempt many times, both individually and in collaboration, each time admitting to our chagrin that we had learned something new but hadn't quite cut away all the brush yet.

Allan's textbook represents a bold attempt to clear the view from theory toward practice. This workbook looks the other way—from practice toward theory. In fact, however, both the theory and the practice evolved together and are still evolving, and you, too, should cut away at the thicket in your mind from both sides. The two books cover much of the same ground but from different perspectives. You need both. Don't risk resources on the information in this book without reading the textbook, and don't manage livestock from the advice in the textbook without mastering the details explained here.

Err in one direction and you risk sticking on dogma that has no relation to reality. Err in the other and you will destroy reality in ways you don't expect.

HOW THIS BOOK WORKS

People attending the holistic resource management courses hear early and often this warning: Never think of the practice as a "grazing system."

Easier said than done.

Most go to the courses to find answers to questions about the land, about weed problems, stocking rates, erosion, or whatever, and in the context of the classroom, the answers seem quite simple. Since the most revolutionary "technical" aspects of holistic management center on the discovery of the beneficial role of herding animals, the tools and guidelines derived from that could easily become the stuff of yet another formula.

Even though the curriculum devotes considerable time to such abstractions as "holism," "applying the holistic model," and "planning" (actually "plan—monitor—control—replan"), the concrete terms "stock density," "herd effect," "cell design," "paddock level," and so forth go down a lot easier among relatives, neighbors, and hands back home.

And in fact "applying the model" often seems unwieldy in the reality that looms outside the classroom. The model's "systematic" approach to thinking things through may make thinking more productive but not a bit easier in the face of real problems—a hiccup in the futures market, a prize cow that tears up fences, an in-law that destroys pickup trucks, drought. Easier to beg off and say, "Oh well, at least we've got more herd effect than before."

No. The last of the old way always beats the first of the new. The first musket couldn't match the best crossbow. The first automobile trailed a good horse. Your first plans may be so far from reality that going back to seat-of-the-pants management appears to make a lot more sense than replanning.

I remember a lady in an adult education class who would not learn to use a ruler. When asked to draw a line through the center of a page, she produced a slash across one corner. "Oh, that's the way it came out with the ruler," she said.

"Well, what about without the ruler?" I asked, knowing that as an expert weaver her native gift for geometry far outstripped mine.

She squinted a second at the paper, then drew a line freehand that missed the mark by a scant sixteenth inch top and bottom. Damn good, but I could still top that with a ruler, despite my inborn lack of talent. So, too, with the analysis and planning processes of holistic management. Learn them, and you will soon beat raw intuition every time.

Financial planning, biological planning, land planning, and monitoring cannot begin, however, without a good grip on some fundamental concepts and a clear idea of your specific priorities. Only after this groundwork can you begin the step-by-step process of generating a plan or, in the case of monitoring, documenting environmental change.

This book is therefore divided into four major parts, each with two sections. The first section in each part covers basic concepts and technique. The second section is a step-by-step procedure for getting your plan down on paper, monitoring it, and adjusting it. In Part IV, "Land Planning," the fundamental structure of basics followed by steps breaks down a bit because land planning is not bound to the strict annual cycle that governs finances, livestock, plants, and crops. Nevertheless, the ultimate objective—getting a comprehensive plan on paper—still holds.

One warning before you begin. Because this book concentrates almost entirely on techniques and procedures, it does not stand on its own. Don't use it until you have familiarized yourself thoroughly with the theory and science of holistic resource management. Read Allan Savory's text *Holistic Resource Management* first, and if possible attend a training course or make contact with someone who has.

ACKNOWLEDGMENTS

This book represents the work of other people to the extent that I functioned more as editor and compiler than as author. The bulk of the ideas came through Allan Savory, out of his long study of the land, and the work of many others besides. Many ideas and refinements have also come out of experiences of the Center for Holistic Resource Management and its staff. The chapter on monitoring is largely a restatement of procedures developed by the Center's education director, Kirk Gadzia. Staffers Champe Green, Naseem Rakha, Roland Kroos, and Ron Moll contributed examples, ideas, comments, and criticisms, as did consultant Ken Williams. Colorado ranchers Reeves and Betsy Brown, Tom Jolly, Gary and Margie Ann Levin, and Miles Keogh did their best to spot examples and numbers that did not seem realistic, and any they didn't catch are not their fault. Jody Butterfield should have her name on the title page for keeping the whole project focused and on track. Medals for patience, forbearance, wisdom, grace under pressure, and other godly virtues go to designer Carolyn Kinsman of TypArt and Island Press editors Barbara Dean and Barbara Youngblood.

———

Dedication

To the memory of Leo Beno, Navajo
range conservationist, killed accidentally
while giving the best of holistic
management and his native skill to the
people of Lesotho—a fine man and
ever-cheerful friend.

———

PART I
FINANCIAL PLANNING

FINANCIAL PLANNING

*T*his book does not begin with land planning, stock density, paddocks, and grazing periods. It begins with financial planning, because money is the ruler. For better or worse it is the ruler even when your goals go far beyond profit. In fact, both the materialist who finds the breath of life in crisp green paper and the poet who would rather live without it might find more peace of mind and freedom of spirit by accepting the notion that money is nothing but a tool. It is certainly not fulfillment and not necessarily even wealth. Holistic thinking would be infinitely harder without the benefit of a tool like money by which to measure progress, though of course money is certainly not the only measurement. By definition holism means dealing with "wholes" in which many elements affect each other simultaneously, but the human brain can't handle everything at once. Since your wallet probably can't either, financial planning is the process through which you will reduce the grand notion of holism to the practical matter of what you do first and how much of it you do, so the whole will come out right.

The very words "financial planning" may remind you uncomfortably of income tax time. Like tax preparation, it is not the form itself but the gathering of information to fill the blanks that presents the greatest challenge. Unlike tax work, however, the end result of financial planning is prosperity and dreams come true.

At the information-gathering stage you will begin to formulate a general strategy based on your analysis of your situation and your goals. The holistic model really comes into its own at this point. Several elements of it in particular will help you keep a perspective on when to use money as the yardstick, when not to, and how to decide priorities.

MASTERING THE BASICS

Goals

As Allan Savory explains in the *Holistic Resource Management* textbook, the whole thing depends upon having a three-part goal that includes quality of life, landscape, and production. As a practical matter you cannot proceed through the steps without keeping this trinity clearly in mind at all times. Ambiguity about your goals may not result in an unworkable plan, but it could very well generate a plan you won't *want* to work (and may even sabotage unconsciously).

Most people who read this book will list "profit from crops and/or livestock" or its bureaucratic equivalent "efficient use of funds" as a production goal. Nevertheless, the person who seeks "maximum profit" and lists "whatever it takes" as his desired quality of life probably has serious delusion, a rotten marriage, a narrow mind, or all of the above.

As money is merely a measurement, so profit is a means to reach other goals. A close family, the creation or preservation of good land, public service, church work, the education of your children, the loyalty to relatives, and many other desires and duties all put demands on profit. If you do not have these things in mind when you plan your commitment of money and labor, you will make a plan that you will inevitably scrap the minute your higher goals demand it.

On the other hand, clarity of goals will enable you to avoid temptations and opportunities of tremendous promise that nevertheless lead in the wrong direction—for you.

"My neighbor is selling out. I could get a great deal on his hay machinery. Why not get it?"

"The government has a cost-share program. Should I participate?"

"My husband just won a trip for two to Maui. Should we go?"

The procedures in this chapter will help you organize a huge amount of complex information about operations that go on simultaneously, but you still have to go one step at a time and put one thing ahead of another. Clear goals make that possible.

It's the same in all professions, even mine.

Guidelines

At this point reread the textbook Chapters 30 to 36 on the testing guidelines: Whole Ecosystem, Weak Link, Cause and Effect, Marginal Reaction, Gross Margin Analysis, Energy/Wealth Source and Use, and Society and Culture. In the financial planning process, all your policies and projects must come up for review through the guidelines as you allocate resources. You may start from a thousand ideas, but before you actually plan action on any of them, you have to evaluate their soundness and set priorities. What enterprises, what investments in land improvements, what training for your staff, and so forth, will you try to carry out with the resources you have? Use the guidelines when you do this! They cut about ninety percent of the confusion out of this task, but unfortunately they don't eliminate it all.

As neat as the guidelines appear in theory, they do overlap a good deal. Sometimes it proves impossible to figure out where one or the other truly applies, but of the seven a few will almost always prove critical in a given case. Some, such as cause and effect, almost always apply. Don't agonize over ambiguity in regard to any one guideline. They all function together like the elements in one of those filters that purifies water through a series of screens, flotations, and catalysts, each of which eliminates one class of contaminant while ignoring the rest.

The textbook covers Whole Ecosystem, Cause and Effect, and Society and Culture well enough to warrant no further practical advice here. The others, however, require some translation into dollars and cents. Let's take these guidelines one by one.

The Weak Link

The training courses pass over the Weak Link guideline very quickly, but it has profound implications for deciding the fundamental question of planning: "What do I do next?" In the classroom the dialogue often goes like this:

Instructor: "If I kept adding tension to this chain, where would it break?"

Class: "At the weakest link."

Instructor: "And if I strengthened that, where would it break?"

Class: "At the next weakest link."

Instructor: "No, at the weakest link."

Class: (Groan)

Although the concept of the weak link may clarify many situations, it particularly applies to long-range planning and accumulation of assets. An immediate situation—say, a unique marketing opportunity—may modify your plans, but in general the best management does not track and exploit random opportunities. You are trying to build an operation and an environment (landscape) that will endure. To do that, you want your major investments of energy and resources to permanently strengthen the weak link.

You will discover many chains and sub-chains on your farm or ranch. A general one is the sunlight energy chain:

Solar Energy Conversion → Product Conversion → Marketing → Reinvestment

Jargon aside, this means that plants make food out of solar energy (just as a cow turns the plants into a saleable product). Then a marketing apparatus converts that product into money, solar dollars, which you can reinvest. In this case the chain makes a circle, because reinvestment is frequently the means for correcting a weak link. All links, of course, ultimately depend on the strength of the people responsible for them. Thus support for training, education, and morale building frequently returns more solar wealth than anything else.

Training courses put great emphasis on the energy conversion link, because ranchers have a tendency to leave that one to God, who unfortunately has seldom interceded on behalf of human neglect. The discovery of the four missing keys to our understanding of the ecosystem, however, passes the buck back to human management. You probably *can* produce a better range that will turn more sunlight into comestible carbohydrates and protein. Nevertheless, this may not be your weak link, so don't assume it.

Here are some things to look for when trying to determine the weak link in the solar chain in a given year:

Livestock Operation

Energy Conversion

Forage shortfall
Paddocks too few to minimize overgrazing
Drainage poor
Species composition poor
Low litter accumulation
High supplement cost

Product Conversion

Unutilized forage
Sufficient forage but—
• Low calving/lambing rate
• Poor gains
• High mortality

Marketing

Low prices
Market resistance due to:
• Insensitivity to demand
• Inadequate research
• Poor product
• Poor sales effort
Ignorance of market mechanisms (futures, etc.)
Unnecessary middlemen

Crop Farming

Energy Conversion

Acreage too small
Inputs too high (fertilizer, etc.)
Poor water management
• Drainage poor
• Overirrigation
Unnecessary single cropping
Wrong choice of crops
Poor crop health

Product Conversion

High damage loss (insects, disease)
Poor germination
High harvest and handling loss
Transport damage

Marketing

Low prices
Market resistance due to:
• Insensitivity to demand
• Inadequate research
• Poor product
• Poor packaging
• Poor sales effort
Ignorance of market mechanisms (futures, etc.)
Unnecessary middlemen

Whatever you determine to be your weak link, you must set in motion a *plan* to strengthen it. This action plan will show up on your financial planning sheets as separate expense columns for training, fencing, advertising, or whatever improves the chain.

The key is whether or not an action directly contributes to the generation of wealth.

Even if purchasing winter feed allows you to carry more animals and thus harvest more during the growing season, it does not directly create solar wealth. Fencing, land acquisition, plantings, or improved drainage, however, do. If product conversion is the weak link, then planning that gives animals access to forage in spite of snow addresses that link; you could raise bison that forage through snow or goats that browse above it. All these solutions have worked somewhere.

Unless the product link is weak, the new swather or engine overhaul does not directly contribute to the wealth generated by the hay field. A new drainage system does because it enhances growth. Therefore, if your equipment will last another year, put the money into tile pipe instead.

Sooner or later, and generally the former, the search may lead you to Pogo's famous discovery: "We have found the enemy and he is us." If your own management isn't good enough to implement your plans, do you stop making them? No. Develop a personal growth plan that improves your management skills and strengthen that link before you waste another cent.

Again, check the Weak Link guideline whenever you have doubts about which course to choose.

Energy/Wealth Source and Use

In the view of holistic management, the foundation of wealth is its intrinsic value—which is rendered from the mineral wealth of the earth or grown by the power of the sun. Money derived from these sources can be termed variously "mineral or petrochemical dollars" and "solar dollars." The exchange value of these dollars nevertheless changes as land values, interest rates, inflation, the price of gold, and so forth fluctuate. Thus "paper dollars" are created and destroyed continuously and, alas, often capriciously.

It is our philosophical position, not shared by all, that ranchers or farmers should measure their success in solar dollars only and rely on the paper ones at their peril. The argument is both moral and practical. The health and fortune of humankind cannot be sustained without the creation of solar wealth, and those engaged in the management of natural resources accept a special responsibility in this regard. On the practical side, the more you can rely on solar wealth, the more you insulate yourself from swings in land and commodity prices, interest rates, and the like. This does not mean ignoring such matters. In fact it demands a particularly nimble and flexible attitude toward them. All borrowing, especially if it is based on the current real estate market, involves paper dollars to some extent, so obviously your position is more stable if you can finance your plans out of solar funds generated by your operation.

The second aspect of this guideline asks whether use is consumptive or cyclical. Is it good only once, or is it durable in that it builds infrastructure for future wealth? As the wording of the guideline—Energy/Wealth Source and Use—indicates, you must apply it to the *source* of energy and wealth that you use and to the *use* of the energy and wealth that you generate. The diesel fuel used in clearing land represents your consumptive use of a nonrenewable energy resource produced elsewhere. Draft horses fed on your own hay and grain might be a positive alternative. Clear cutting timber to make a loan payment is a consumptive use of your solar wealth if it destroys the productivity of the soil. Proper forestry or conversion into sustainable pasture might pass this guideline.

It is particularly helpful to apply this guideline as you weigh ideas for strengthening the weak link in that solar chain of new wealth generation. Of dozens of ways to proceed, some will represent a better use of energy and wealth than others. At the product conversion level, many rationalize heavy grain feeding to push livestock to a higher market class. Though the cash return may appear too good to refuse, the feed typically represents a consumptive use of soil and fossil energy. Finding a different market might reduce the need or even eliminate it altogether.

Marginal Reaction

To some degree the Marginal Reaction guideline is the accountant's equivalent of the Weak Link test, but it applies in many nonquantifiable situations as well. It also applies at a level of greater detail.

If you followed the guideline perfectly you would make your investments one dollar at a time, asking for each one, "Where will this move me furthest toward my goals?" Each dollar goes where it will yield most, and this changes whenever diminishing returns on one investment drop below what the next dollar might return somewhere else.

The textbook illustrates this process by a clear example using dollars and percentages. The only pitfall in practice is an expectation that you can quantify every situation. You can't. Since your goals can't be quantified, you will have to compare apples and oranges.

Suppose energy conversion is your weak link. You can buy land, lease land, develop water on land not now accessible, reseed old land, hire herders or build fence, or retrain your staff. Which will have the greatest marginal reaction?

Advertising may strengthen a weak marketing link if you have to establish a market. After a point, however, extra publicity has

Debt, Overhead, Risk, and Scale

The key to survival in a world dominated by paper dollars is remaining nimble—able to shift quickly from one enterprise to another or even to sit out a bad market. In no industry is this more true than agriculture, where markets, weather, land prices, and input costs fluctuate drastically and without warning. The two graphs illustrate how costs (fixed and variable),

machine (breed, land, irrigation system, fence) and your cost of production will go down, and you'll realize a bigger profit." As you can see from the graphs, however, this argument only becomes true when production passes 10 units. At anything less, Case II does better. If this "equality point" represents a fairly optimistic level of production, the Case I producer suf-

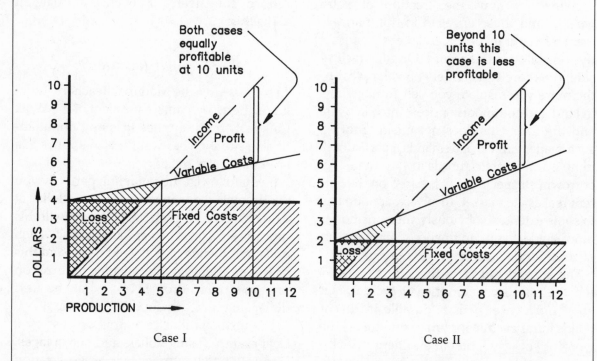

Case I

Case II

risk, and scale of production relate to each other. If you study them long enough to fix them in your mind, they'll help you think through any number of investment decisions.

Graphs in Cases I and II show two ways to approach the same enterprise. In Case I the fixed costs are twice as high as in Case II, but the variable costs are half as much, as indicated by a variable costs line half as steep (1/5 dollar per unit against 2/5 dollar per unit). The argument for Case I runs thus: "Invest in this

fers a lot of stress when fate deals him just an average run of cards. He has to pay the high ante and bet heavily just to stay in the game.

The pressure becomes even greater if the fixed costs reflect investments secured by paper values such as land appraisals, which typically also lose value when general conditions are bad. On top of that, risk exposure, as represented by the shaded "loss" area on the two graphs, is vastly greater for Case I and its predictable-as-death high costs.

The two graphs in fact give Case I the benefit of the doubt because they do show a trade-off of lower variable costs for higher fixed (capital) costs. If the Case II operator could find a way to reduce the slope of his variable costs line to equal Case I, he would of course enjoy higher profit at all levels of production and less risk exposure as well—a nice trick if you can do it.

In practice, many additions to fixed costs do reduce variable costs. But they almost always raise the threshold of profitable production, increase risk, and cut flexibility. Before you borrow and invest, compute that higher threshold carefully and reflect a while on what it means for your finances, your peace of mind, your work load, and your general quality of life.

If you can cut the slope of your variable costs line without adding fixed costs, do that first. Thus management improvements that allow you to increase stocking rates or reduce supplements or labor give you more flexibility than assuming mortgage payments on additional land. When you do compare capital investments—such as expanding through land purchase or fencing to improve your stocking rate—be sure to compute the precise effect on the slope of that variable costs line.

Whether or not a small family farm can stay in the game turns on the balance of fixed and variable expenses and scale of production, and the deck is sometimes stacked. The most significant economies of scale in agriculture exist in the supporting industries—machinery, chemicals, processing, and advertising—not on the farm itself. John Deere profits most by making a big volume of very big combines. Cargill profits from handling grain by the trainload. Since ultimately their money comes from your solar dollars, they'd like to persuade or force you to fit their pattern.

The producer can win by holding onto solar dollars through creative marketing and keeping production flexible. You can plant a new crop much easier than John Deere can open a new production line. If getting bigger means debt and inflexible capital investments, it will force you to play by rules written by multinational, high-volume industries, and they will win at your expense.

only a short-term effect. Then storage facilities to gain market flexibility or better breeding stock to improve quality might have greater marginal reaction. They might enable you to enter new markets and stop wearing yourself out holding your share of a limited one. You would thereby retain more of the solar wealth generated on your land.

Perhaps none of these strategies generates an acceptable marginal reaction in relation to your quality of life goal. More carrying capacity or better sales might increase your income, but your daughter needs a loan to buy the place next door. Maybe that will pay back more in joy than the same money ever could in beef.

The matter is further complicated by the old truth that time is money (and often much more than money). You must consider the marginal reaction of time. Will the hour spent riding fence return more toward the fulfillment of your goals than the hour spent planning? How about spending the same hour with your family?

Think marginal reaction. You are building a life and a landscape and a source of solar wealth that will sustain itself as far down the road as you dare look. You have to figure how far down that road each dollar will send you.

Gross Margin Analysis

Simply stated, gross margin analysis is a technique that separates fixed costs (defined here as costs you have regardless of what or how much you produce) from the costs directly linked to production. It enables you to compare many enterprises or combinations of enterprises. Besides expenses normally termed overhead, fixed costs can include other expenses that have already been incurred. For example, fertilizer—normally thought of as a variable cost—can be fixed if already bought and you're making a new decision about crops.

This approach, developed by David Wallace, an English economist, respects the fact that at any point many costs such as living expenses, debt payments, and full-time labor exist no matter what you produce. So they should not be considered in judging the potential of an enterprise you undertake tomorrow.

Many people do not make this distinction when analyzing the sources of profit and loss. They look only at the bottom line—which tells them little or nothing about which elements of an operation contribute most to covering overheads and ultimately whether or not those overheads are justified. Even when different enterprises in an operation are analyzed separately, it's common to apportion some fixed costs among them. But this obscures the real contribution of that enterprise. People unused to this concept often find the distinction between fixed and variable ex-penses confusing and want a list. Again, the decision is subjective. Any expense that you can't avoid in the planning period regardless of what you plan is a fixed expense. Any expense that derives from your plan is variable.

Failure to see the difference can lead to disastrous decisions. The two graphs shown below, for example, illustrate the large-scale ruin of dairy operations in the mountains of western North Carolina. Beginning in the 1960s industry experts encouraged local dairymen to build continuous milking parlors in association with feedlots supplied by intensive crop production or purchased feed. In theory this would produce more profit per cow—and virtually no end to the number of cows you could have.

The left-hand graph shows how traditional accounting justified this strategy. Cost per cow is enormous, but so is production, so gross profit per animal looks good enough

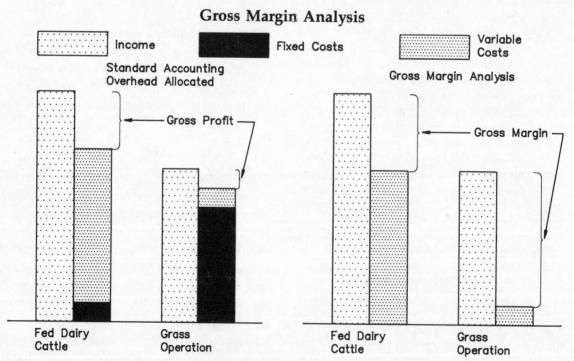

Gross Margin Analysis

The left-hand columns compare profits when fixed land costs are allocated according to the amount of land used. The right-hand columns compare gross margin when land costs, which have to be paid anyway, are left out.

to justify the large capital investment in the new equipment and feed purchase or production. Everything "cash flows." The century-old practice of grazing cows on steep mountain pastures looks pretty poor in comparison.

At the time of decision, however, all investments in the continuous milking scheme were *variable* costs! These included not only such obviously variable costs as feed, vet bills, and semen to produce high-production cows. An annual cost per cow over the projected life of the new facilities was also a variable cost—even though it would become a fixed cost after the investment was made.

The most significant fixed costs were mortgage payments on land and taxes, which were high because booming second-home development had inflated values. Since only a fraction of the mountain land was suitable for crops and feedlots, little of this cost was assigned to the intensive milking systems, even though it had to be paid. Meanwhile huge areas of pasture, laboriously cleared in past decades, went out of production under the new system and in that nonbrittle environment quickly succeeded to forest. There was no going back. Those who could sell land to developers survived financially, but the dairy herds disappeared.

Gross margin analysis would have revealed the problem before the fatal step, as the second graph shows. If variable costs alone are considered, it becomes obvious that each cow on the old mountain grass went much farther toward covering the cost of the land and incurred relatively little "per head" expense compared to the cow on feed. Even when the high taxes ate up profit, die-hard farmers could often supplement their income enough to hold onto their property because the whole farm produced. Gross margin per acre remained healthy, and other inputs were low. The story might have turned out quite differently on flat, fertile land. But in that case

a gross margin analysis might well have led to truck farming instead of intensive feed crops and Holsteins.

Gross margin analysis tends to influence your planning in several key ways:

1. You can be far more nimble in response to paper dollar changes. You will know when buying land makes sense, when sheep will do better than cattle, when yearlings beat calves, when to grow hay and when to buy it.

2. You will make much better use of the land, equipment, and labor you already have. The common practice masks the expense of idle assets. Gross margin analysis shows when you're better off making smarter use of what you have than making "cost-effective" new investments.

3. It makes you extremely aware of overhead costs. In the long run, none of these costs is really fixed at all. You can sell the ranch itself if you have to. Short of that you can bias your planning to cut overhead and get rid of inefficient assets. Thus you can cut your exposure to the dangers of paper dollars—the great trap of American agriculture.

The following simplified version of a real situation illustrates the technique of preparing a gross margin analysis. The numbers are in South African rand, not dollars, and the relationship between income and animal numbers does not transfer either, so don't expect the conclusions to relate to your operation. The method, however, will. In this particular case two brothers differed as radically as Cain and Abel on the subject of sheep versus registered cattle. The analysis answered their pressing economic question but not, alas, the human one.

The income and expense account shown here has had a lot of detail boiled out of it, but it still includes many items that do not fiqure in the gross margin analysis. These represent items that either have nothing to do with the enterprises in question or are not "variable" in the sense that you will not get rid of them in the near future regardless of your conclusions about the various enterprises. Note that the analysis side has some categories such as "pasture, silage, and hay" that do not rate a line on the traditional account but are derived from several of the expense headings.

To compare the livestock options effectively, gross margin analysis results must be reduced to some common and comparable base—dollars per head per year, dollars per acre per year, dollars per standard animal unit per year, dollars returned per dollar invested per year, and so forth.

In this case, we are given

1,500 sheep
400 commercial cattle
150 registered cattle

The margin per head (in rand) would be

$$\frac{R11,580}{1,500} = R7.72/sheep$$

Comparing Enterprises

Income & Expense Account

Income

Livestock operations	
Trading	42,592
Herd value change	(359)
Produce	
Wool	5,580
Dairy	714
Timber	72
Miscellaneous	
Show prizes	237*
Equipment sold	1,360
Bonus and rebates	576
Dividends	3
Total Income	**R50,775**

Expenses

Feed, salt, etc.	12,113
Crops (seed and chemicals)	10,033
Vehicle and equipment operation	9,660
Repairs (fence and machines)	1,507
Labor	
Seasonal	1,888
Full-time	14,152
Veterinary and A.I.	3,412
Shearing	730
Registered stock fees	
Electricity	1,060
Insurance	2,744
Administration	1,997
Interest	5,426
Miscellaneous	189
Total Expense	**R65,521**
Profit (Loss)	**(R14,746)**

Gross Margin Analysis

	Sheep	Grade beef	Registered cattle
	23,230	9,454	9,908
	(2,292)	6,900	(4,967)
	5,580	—	237*
	R26,518	R16,354	R5,178

Gross Income from Enterprises

Less Variable Costs

Sheep	Grade beef	Registered cattle
3,294	4,317	4,452
9,414 pasture		
	3,600 silage	1,772
	4,871 hay	2,400
1,500	1,239	673
730		
		610
R14,938	R14,027	R9,907
R11,580	R2,327	(R4,229)

Gross Margin

*Note: Show prizes relates to registered cattle.

$$\frac{R2,327}{400} = R5.82/\text{commercial cow}$$

$$\frac{(R4,229)}{150} = R28.20/\text{registered cow (loss)}$$

If the stocking rate reflects the U.S. Bureau of Land Management ratio of five sheep to one cow unit, then the margin difference per unit of land is drastic indeed. Sheep return R38.60 per animal unit in this particular case; registered cows lose R28.20.

Gross margin analysis helps in planning and monitoring, but it doesn't replace the old books because it does not consider all your expenses. Use it for three purposes:

1. To compare enterprises as in the example

2. To see how different conditions such as price changes might affect your enterprise

3. To weigh the effect of diverting assets from one enterprise to another

Comparing Scenarios

Gross margin analysis can be used to assess the impact of future events or policies. In the example below, "poor," "average," and "good"

Example of Gross Margin per 100 Cows

Gross Income	Poor	Average	Good	
Weaning %	85%	85%	85%	
Price/lb.	$ .50	$ .85	$ 1.00	
Weaner wt/lbs.	500 lbs.	500 lbs.	500 lbs.	
Weaner gross income	$21,250.00	$36,125.00	$42,450.00	Least
Cull cows 20% (-2% mortality)	18%	18%	18%	farmer
Cull cow price	$ 330.00	$ 475.00	$ 550.00	control:
Cull cow gross income	$ 5,940.00	$ 8,550.00	$ 9,950.00	To keep the arithmetic
Gross Income	**$27,190.00**	**$44,675.00**	**$52,400.00**	simple, vary only the
				factor least
Variable Costs				in your
Replace heifers 20 @ $400;$500;$550	$ 8,000.00	$10,000.00	$11,000.00	control.
Veterinarian expense @ $10/(cow)	$ 1,000.00	$ 1,000.00	$ 1,000.00	
Supplies (3 lbs./day for 3.5 months)	$ 3,500.00	$ 3,500.00	$ 3,500.00	
(330 lbs. @ $160/ton)				
Interest @ 10% on average value $500	$ 5,000.00	$ 5,000.00	$ 5,000.00	
Bull cost $12/(cow)	$ 1,200.00	$ 1,200.00	$ 1,200.00	
Shipping	—	—	—	
Variable Costs	**$18,700.00**	**$20,700.00**	**$21,700.00**	
Gross Margin (per 100 cows)	**$ 8,490.00**	**$23,975.00**	**$30,700.00**	

Calculating bull cost per 100 cows
Purchase @ $1,500 Sell @ $1,000 Life 5 years   Depreciation $100.00

Interest 10% $\dfrac{\$1,500 + \$1,000}{2} \times 10\% = \$125$ $125.00

Veterinarian and feed $50   50.00

Miscellaneous   25.00

        $300.00

1 bull to 25 cows $\dfrac{\$300}{\$25} = \$12/\text{cow}$

refer to price only. You could use the same method to find out instantly how changes in supplemental feed prices, labor policy, weaning percentages, average gain, and other factors would affect your margin.

Comparing the Use of Assets

Gross margin analysis allows you to estimate the effect of changing an asset from one use to another. This could be land, machinery, buildings, cash, or whatever. Below we dissect the question of whether to use a piece of bottomland for alfalfa or as pasture for stockers. The numbers are simply pulled out

ALFALFA — 100 acres

Income

3.5 ton/acre @ $75/ton	$26,250.00

Expenses

Custom cut, bale, and stack @ 1.20/bale	13,860.00
Seed and planting $5,000/5 years	1,000.00
Chemicals	1,950.00
Application costs	1,050.00
Total	$17,860.00
Gross Margin	$ 8,390.00
Gross Margin/Acre	**$ 83.90**

STOCKERS — 100 @ 1 steer/acre

Income

750 lb. steers @ $.62/lb.	$46,500.00

Expenses

500 lb. steers @ $.70/lb.	35,000.00
Salt and minerals $1/head/month	500.00
Shipping	1,000.00
Fence and miscellaneous	300.00
Interest	1,000.00
Total	$37,800.00
Gross Margin	$ 8,700.00
Gross Margin/Acre	**$ 87.00**

of a hat to show the technique. Don't judge their plausibility.

Notice that the two options produce very similar bottom lines, but the process forces you to ask (and answer) many important questions:

- Which choice is most risky?
- Would your own machinery make hay cheaper?
- Could you raise your own stockers?
- Could you finance stockers yourself?
- Could you increase the stocking rate?
- Could you increase yield?
- Could you cut chemical costs?
- Which gives most flexibility in the future?

Gross Margin Per Unit

If the mathematics of reducing gross margin to a unit of something tends to slip your mind, remember:

$/head, $/acre, $ margin/$ invested, miles/hour, pounds/square inch, and so on

are written like fractions, because they are. The top gets divided by the bottom.

If 100 cows produce a gross margin of $24,000, then

$$GM/cow = \frac{\$24,000}{100} = \$240/cow$$

If the herd uses 800 acres, then

$$GM/acre = \frac{\$24,000}{800} = \$30/acre$$

If the herd would cost $50,000 to replace, then

$$GM/\$ \text{ invested} = \frac{\$24,000}{\$50,000} = 0.48$$

That is commonly said as "forty-eight cents on the dollar" or 48%.

The Danger of Gross Margin Analysis

Gross margin analysis tells you what each enterprise contributes toward overhead expenses and possible profit. Only the final financial plan, however, really tells you if the enterprises you have chosen will together pay all the overhead expenses and actually return a profit. In any case, you must carry out a full financial plan to make sure that all your overheads are in fact paid. Like any other technical gimmick, gross margin analysis doesn't replace common sense.

Especially in crop farming, a particular crop may promise a high gross margin, but fail all the other testing guidelines because of its use, its nature, or its associated technology. The farmer whose hand calculator proves that cotton yields more bucks per acre than alfalfa on the strength of a gross margin analysis could easily destroy his land by monocropping cotton year in and year out. The family that made a killing in strawberries this year should not necessarily cut down an apple orchard to double the berry crop.

Diversity is good in itself. It reflects a higher, more stable level of succession in nature and a hedge against changing conditions in the free market. Gross margin analysis used with the remaining six testing guidelines avoids the trap of cash returns at the expense of true biological capital and, eventually, quality of life.

The simple form of gross margin analysis described here has one other obvious limitation as well. It does not account for time. You cannot of course schedule crops or other operations that demand more labor, equipment, and capital on a given day than you can supply.

Use gross margin analysis, but think and work to achieve both a form of cropping and a selection of input items that pass the remaining six guideline tests.

Good Ideas
Brainstorming

The practice of holistic resource management involves a number of formal activities—financial, biological, and land planning, monitoring, and controlling. Embedded in all of them, though, is the informal ritual of brainstorming.

Some people wince at the word "ritual" because it sounds like church, but in fact people use rituals all the time to prepare for very ordinary activities. Think of warming up for an athletic event, dressing for a dance, or the daily routines that carry you from waking up to starting work.

The brainstorming ritual is designed to open the minds of planners to new ideas. "Serious" people sometimes shy away from it, but research has shown that the most original and fruitful thinking occurs during moments of humor and playful competition. Since the very phrase "financial planning" tends to produce the opposite mood, a ritual helps.

Before grinding your brains for new ways to make money, cut costs, streamline calving, build fence, or lay pipe, try this one to get everyone's mind in a playful, creative state. The rules are simple:

1. Gather everyone who might have an interest in the task at hand and some people who don't.

2. Divide the people into groups of five or more (eight to ten maximum), and let each group appoint a recorder.

3. Announce a timed competition (about 5 minutes) for the longest list of ways to solve a lighthearted problem. Any idea will do, no matter how crazy. No judgment will be made. Only the number counts. Pour out ideas. Do not stop to talk or discuss.

4. Start the timer and at the end of the time read the lists.

5. Now put the serious problem on the table and let the groups compete in the same way (for 10 minutes), listing any solution that pops into their head, no matter how fantastic.

6. From these lists pick out the ideas that have potential and develop them.

Here are some typical challenges for the opening competitions:

• What uses could you find for . . . (some simple object)?

• How would you get an interview with . . . ?

• How might you deliver a proposal of marriage to . . . ?

To emphasize the point that open speculation often produces more than reworking old ground, Allan Savory tells the following story: "While I was touring a ranch in Africa, my host took me to a particularly rough and inaccessible section that he said had no use whatever. It was steep, waterless, full of predators, and a thorough headache to a stockman, but it was beautiful. 'Why not put some cabins down there where couples from the city could hide out for a little romance?' I asked. The rancher, a man of righteous constitution, took great exception to the hint of impropriety, but the notion that natural beauty and tranquility had cash value in themselves stuck. He built the cabins, and within a few years they produced more revenue than the rest of his operation combined. The moral is that pounds of beef per acre may be only one of a million ways to assess the potential of land and the value of its production."

In a similar vein it might be remembered that when Ebenezer Bryce first looked down on Bryce Canyon in Utah, he could only remark: "That's a hell of a place to lose a sheep." Yet how many millions of Americans have driven how many millions of miles for the same view?

Net Managerial Income

Paying employees a bonus when the enterprise turns a profit is an old idea. Unfortunately its potential for inspiring people to make their work more productive usually remains unfulfilled—the big bottom-line figure is so remote from daily decision-making that most workers never really know how they fit in.

Motivation increases impressively if even lower-level supervisors can see what they contribute to profit. When managers participate in financial results, net managerial income (NMI) provides a fair measure on which to base their compensation. NMI is the difference between gross income and gross expenses, both under the direct control of a management sector. This could be a foreman who has total responsibility for a band of sheep. It could be the managers of farming and livestock divisions in a large operation.

NMI accounting might also recognize the frequently made distinction between "owner control" and "management control." In this case the owner oversees such matters as mineral leases, real estate transactions, and futures trading, while a separate management team runs agricultural production.

NMI is the simplest index of managerial effectiveness. Setting up the columns on your planning chart (described further on) to show it graphically month by month can provide a major staff incentive—especially if NMI is connected to financial rewards and promotions.

NMI has obvious advantages because it creates incentives for managers to cut costs as well as raise income and keeps them aware of where the costs are. Unfortunately, though, there is no universal formula for relating NMI to actual staff payments, because not all sub-managers control equal shares of income and cost amounts. For the manager of a cattle enterprise, for instance, the main cost may be fixed land payments over which he has no control whatever. If you take land payments out of the cattle account, however, you'll show an unrealistic net income. In contrast, the manager of a chicken facility may have great control over the big expense of that enterprise—the feed.

You can compensate for such inequities, but often the exercise will cause you to redefine jobs. If, for instance, the person who manages field crops has no responsibility for machinery costs and maintenance, you might want to make that part of the NMI account and grant the authority necessary for controlling those costs. Such reorganization always runs the risk of upsetting entrenched interests. Be sensitive, proceed openly, and make every effort to build trust in advance.

Emergency Decisions

Moments will arise when everything seems to be going haywire. In gathering information for planning you may well uncover some unmonitored nook of your operation where everything is already haywire. Just as the brainstorming ritual will help shake loose new ideas, there's an emergency decision ritual that will help you stave off panic.

Allan Savory, who teaches this technique in his courses, credits this one to British military training, where it's called a "Simple Appreciation." Soldiers have long been taught to substitute routines for panic, so when bullets fly they duck and take aim before fear makes them stand up and run. It seems they also learn a routine for thinking when no routine applies:

1. Define your objective in one paragraph. What result do you want?

2. List all aspects of the problem and any information bearing on it. Do this in no order. Just put all ideas on paper.

3. Outline several courses of action (certainly more than one).

4. Pick the best plan.

Just making yourself stop and go through these steps will clear your head and focus your imagination.

Worksheets

Before combining all your ideas into a single grand plan, you must work out the details of each aspect in terms of labor, quantities, expense, and most of all *time*. Time figures critically in all decisions. Work loads that coincide may require more labor. Expenses that come before your products are marketed may require credit. Construction not completed on time can cost you an entire crop.

The handiest worksheet form developed so far has columns for the months of the year (you can start with any month to encompass the natural cycle of a task) and rows for different categories of expense, work, income, decisions, or whatever. The possible uses are far too varied to illustrate here. Your whole operation may well require dozens of worksheets and many drafts. You can use final drafts not only for assembling the master plan but also as general work schedules, inventory and budget guides, and the basis for all ongoing monitoring and controlling. The examples that follow show something of the worksheet's versatility.

The worksheet on page 18 shows the "biological year" of a cow herd. (Appendix A contains a blank example you can copy.)

- It helps in designing a culling policy or when changing herd size.

- It shows animals in each class for reckoning supplements, grazing, and the like.

- It helps you to plan sales policy, holding late calves, breeding back open heifers, and so forth.

WORKSHEET

CENTER FOR HOLISTIC RESOURCE MANAGEMENT
BIOLOGICAL YEAR — COMMERCIAL CATTLE   Date 1986   Ranch RIVER BEND   WORK SHEET NO. 1

Detail	January	February	March	April	May	June	July	August	September	October	November	December	Total
	CALVING	BULL YEARLING HEIFERS		PULLING					‹WEAN				
MATURE COWS				(38+→)									
HEIFERS 2	(36)	(37)	(38)	(27)	(28)	(29)	(30)	(31)	(32)	(33)	(34)	(35)	
HEIFERS 1	(24)	(25)	(26)	(15)	(16)	(17)	(18)	(19)	(20)	(21)	(22)	(23)	
HEIFERS	(12)	(13)	(14)							(9)	(10)	(11)	
FEM. CALVES									(8)	(9)	(10) SELL CULLS		
MALE CALVES									(8)	(9)	(10) SELL		
CULLS									SELL CULL BULLS, COWS AND HEIFERS				
BULLS	BUY												
Total													

The top line shows major events in the biological plan—calving, weaning, and such. The long arrows show when animals progress from one class to another, or go to sale. Numbers in brackets show average age.

EXAMPLE

Alfalfa Production and Sales Plan

Here's an alfalfa production and sales schedule. It represents a strategy of spreading the income around the year. Later in the planning, this could be modified as the cash flow situation becomes evident.

WORKSHEET

CENTER FOR HOLISTIC RESOURCE MANAGEMENT
160 ACRES ALFALFA   Date 1986   Ranch RIVER BEND   WORK SHEET NO. 2

Detail	January	February	March	April	May	June	July	August	September	October	November	December	Total
										480 TONS KEPT FOR USE ON THE RANCH			
HARVEST AV. 3 TON/ACRE						480 TONS	480 TONS	480 TONS				1,440 TONS	
SALES		SELL 120 TONS				SELL 480 TONS	SELL 240 TONS					SELL 120 TONS	
ESTIMATED PRICE		$100/TON				$80/TON	$80/TON					$90/TO	
CASH INCOME		$12,000				$38,400	$19,200					$10,800	$80,400
Total													

EXAMPLE

Gasoline Use and Purchase Projections

This worksheet shows gasoline purchases for ranch vehicles. It projects consumption by various employees, inventories of gas bought in bulk, cost and timing of bulk purchases, and monthly off-ranch purchases.

WORKSHEET

CENTER FOR HOLISTIC RESOURCE MANAGEMENT   Date_____   Ranch_____   WORK SHEET NO. _____

Detail	January	February	March	April	May	June	July	August	September	October	November	December	Total
	←FEED AND CALVES→		←BULLING→						‹WEAN	SALE›	←FEED→		
MANAGER JEEP	200	250	250	250	250	250	250	250	250	400	100	100	2,800
HERRERA PICKUP MI.	200	200	150	50	50	50	50	50	50	50	200	200	1,300
FOURNIER PICKUP MI.	350	350	250	250	60	60	60	60	60	60	350	350	2,260
SMITH PICKUP MI.	100	100	100	300	300	300	300	500	500	500	100	100	3,200
TOTAL MILES	850	900	750	850	660	660	660	860	860	1010	750	750	9,560 MI.
GALS. GAS @ 12 MI./GAL.	71 G.	75 G.	63 G.	71 G.	55 G.	55 G.	55 G.	72 G.	72 G.	84 G.	63 G.	63 G.	797 G.
HERRERA MI. BIKE GALS.	—	—	100 1.2 G.	100 1.2 G.	300 3.6 G.	300 3.6 G.	300 3.6 G.	300 3.6 G.	300 3.6 G.	300 3.6 G.	—	—	
FOURNIER MI. BIKE GALS.	—	—	150 1.8 G.	150 1.8 G.	400 4.8 G.	400 4.8 G.	400 4.8 G.	400 4.8 G.	400 4.8 G.	400 4.8 G.	—	—	
TOTAL GALS.	71 G.	75 G.	66 G.	74 G.	63 G.	63 G.	63 G.	80 G.	80 G.	92 G.	63 G.	63 G.	853 G.
1/3 OFF RANCH @ 95¢/GAL.	$22	$24	$21	$23	$20	$20	$20	$25	$25	$29	$20	$20	$269
BULK GAS @ 80¢/GAL.	500 G. $400								400 G. $320				
RANCH USE GALLONS	47 G.	50 G.	44 G.	49 G.	42 G.	42 G.	42 G.	53 G.	53 G.	61 G.	42 G.	42 G.	
Total													

The Livestock Production Plan

This form is a glorified worksheet for livestock operations. Fill it out with the help of a "biological year" worksheet. The form has three sections and can be used to show three years or three divisions of a given year. Use a separate sheet for each livestock enterprise (sheep, commercial cattle, llamas, and so on).

The Livestock Production Plan tells you when your expenses and income will occur and predicts the effect of breeding rates, culling policies, and the like over a long period. For example, you can tell at once what effect a 5% drop in conception rate this year will have on your herd size in 3 years. It can be computerized so you can answer questions like that in microseconds.

CENTER FOR HOLISTIC RESOURCE MANAGEMENT

LIVESTOCK PRODUCTION PLAN

RANCH __RIVER BEND RANCH__
ENTERPRISE __COMMERCIAL CATTLE__   DATE OF PLAN _____
REMARKS __3 YEAR PROJECTION (CULLING UNPRODUCTIVE HEIFERS, 20% OF COWS, 10% OF HEIFER CALVES)__

Opening inventory

A. CLASS OF STOCK	B. BIRTH EST %	C./D. OPEN NO.
1. BULLS		30
2. COWS	90	500
3. HEIFERS 2	70	200
4. HEIFERS 1	90	225
5. FEM. CALVES		240
6. MALE CALVES		0
13. TOTAL HEAD		1,195

YEAR OR MONTHS 1986

CLASS OF STOCK	BIRTHS	BUY Mth	BUY No	TRANSFERS IN	TRANSFERS OUT	DEATH	SALE %	SALE No	SALE Mth	CLOSE & OPEN NO.	AGE
BULLS		10	10	2	1			10	10	30	30
COWS				200		14	2	138	11	548	
HEIFERS 2				225	200	4	2	66	11	155	33
HEIFERS 1				240	225	5	2	24	11	211	21
FEM. CALVES	396	3			240	16	4	38	11	342	
MALE CALVES	396	3				16	4	380	11	0	9
TOTAL HEAD	792		10	665	665	55		646		1,286	

YEAR OR MONTHS 1987

CLASS OF STOCK	BIRTHS	BUY Mth	BUY No	TRANSFERS IN	TRANSFERS OUT	DEATH	SALE %	SALE No	SALE Mth	CLOSE & OPEN NO.	AGE
BULLS		10	12	2	1			12	10	30	30
COWS				155		14	2	138	11	551	
HEIFERS 2				211	155	4	2	62	11	145	
HEIFERS 1				342	211	7	2	34	11	301	
FEM. CALVES	435	3			342	16	4	38	11	342	
MALE CALVES	435	3				16	4	380	11	0	
TOTAL HEAD	870		12	708	708	57		664		1,369	

YEAR OR MONTHS 1988

CLASS OF STOCK	BIRTHS	BUY Mth	BUY No	TRANSFERS IN	TRANSFERS OUT	DEATH	SALE %	SALE No	SALE Mth	CLOSE NO.	AGE
BULLS		10	13	2	1			13	10	30	30
COWS				155		14	2	139	11	553	
HEIFERS 2				201	155	6	2	89	11	96	
HEIFERS 1				342	201	7	2	34	11	401	
FEM. CALVES	435	3			342	17	4	42	11	376	
MALE CALVES	435	3				17	4	418	11	0	
TOTAL HEAD	870		13	698	698	61		735		1,459	

NOTE: RECORD YOUR ESTIMATES OF PERCENTAGE ACTUAL BIRTHS FOR THE VARIOUS AGE CLASSES OF BRED FEMALES IN THEIR ROWS IN COLUMN (B) ABOVE.

ANALYSIS OF PLANNED SALES AND PURCHASES

1986

	BULLS	COWS	STEERS	HEIFERS H2	HEIFERS H1
14. CLASS OF STOCK					
15. NO. SOLD/MONTH OF SALE	10 / 10	138 / 11	380 / 11	66 / 11	24 / 11
16.					
17. AVERAGE LIVE WEIGHT	1,500	950	550	900	800
18. MEAT PRICE PER LB.	60¢	70¢	75¢	70¢	70¢
19. INCOME PER ANIMAL	900	665	412	630	560
20. WOOL/HAIR WT./MTH. OF SALE					
21. WOOL/HAIR PRICE/MTH. OF SALE					
22. WOOL/HAIR INCOME					
23. PLANNED GROSS INCOME	9,000	91,770	156,560	41,580	13,440
24. NUMBER PLANNED TO BUY	10				
25. ESTIMATED PRICE/ANIMAL	660				
26. TOTAL COST AND MONTH	6,600 / 2				

1987

	BULLS	COWS	STEERS	HEIFERS H2	HEIFERS H1
14. CLASS OF STOCK					
15. NO. SOLD/MONTH OF SALE	12 / 10	138 / 11	380 / 11	62 / 11	34 / 11
17. AVERAGE LIVE WEIGHT	1,500	950	550	900	800
18. MEAT PRICE PER LB.	60¢	70¢	75¢	70¢	70¢
19. INCOME PER ANIMAL	900	665	412	630	560
23. PLANNED GROSS INCOME	10,800	91,105	156,560	39,060	19,040
24. NUMBER PLANNED TO BUY	12				
25. ESTIMATED PRICE/ANIMAL	660				
26. TOTAL COST AND MONTH	7,920				

1988

	BULLS	COWS	STEERS	HEIFERS H2	HEIFERS H1
14. CLASS OF STOCK					
15. NO. SOLD/MONTH OF SALE	13 / 10	139 / 11	418 / 11	89 / 11	42 / 11
19. INCOME PER ANIMAL	900	665	412	630	337
23. PLANNED GROSS INCOME	11,700	91,105	172,216	56,070	14,154
24. NUMBER PLANNED TO BUY	13				
25. ESTIMATED PRICE/ANIMAL	660				
26. TOTAL COST AND MONTH	8,580				

This production plan tracks a herd for 3 years according to the biological year described in the worksheet on page 18. It reflects a policy of culling 20% of mature cows, 10% of female calves, and all unproductive heifers (10% of Heifers 1 and 30% of Heifers 2). Study the numbers with a calculator in hand until you get the hang of how the numbers develop.

Fill out the columns in order from the left:

A. List the same classes of livestock used in the worksheet.

B. Estimate birthrates.

C. Enter the beginning number in each class.

D. Enter the average age in months.

E. Calculate births from all classes using rates from column B.

F. Enter month of average calf.

G. List purchases in any class.

H. Enter month of purchase.

Class transfers. Enter from column C the number moving in from the class below and out to the class above. Those moving out from one class must equal those moving to the next.

I & J. Account for death losses. Estimate the rate in column J and calculate the number in column I.

K & L. Show sales in each class and month of sale.

M & N. Show closing numbers and average age.

At the bottom of the Livestock Production Plan is a space for analysis. Fill it out as follows:

14-16. Enter classes from column A, numbers sold, and month of sale. Use row 16 if animals are sold in two groups.

17-19. Estimate weight, sale price/lb., and income/head.

20-22. Use these rows for wool/hair sales.

23. Enter gross income and month received.

24-26. Enter purchase estimates and expense.

CREATING YOUR PLAN

You've done your gross margin analysis of various enterprises. You know in your heart what's right for you. You have some dollars in the bank, some debts to pay, and Christmas is coming. Time to plough paper, sow ink, and reap dreams. As the English poet V. Sackville-West put it in verse:

Under the double spell of night and frost
Within the yeoman's kitchen scheme
The year revolves its immemorial prose.
He reckons labor, reckons too the cost;
Mates up his beasts, and sees his calf-
*   run teem.*

What follows here is the series of steps that Allan Savory worked out for creating and using a total farm or ranch plan. It differs fundamentally from the usual cash flow planning that agriculture had adopted from other industries—which tends to determine priorities in terms of cash and cash return only.

Planning the Planning

Most cost accounting procedures treat all capital items as "wealth generating" so long as they can be shown mathematically to return enough revenue to pay for themselves and provide an acceptable margin. This approach consistently leads to failure, though, because wealth-generating potential also includes such noncash items as soil productivity, water quality, and training.

The common practice also fails to rank investments in relationship to larger goals. Again, anything goes as long as it returns enough dollars. Here we try to discover what will return enough in terms of progress toward goals that may not be quantifiable—say, maintaining a diversity of species. This makes a particular difference in funding capital investments. Instead of putting your money into many projects simultaneously, you'll probably find yourself tackling one at a time but more intensely, as not very many are likely to pass the testing guidelines in a given year.

The planning process outlined here represents many years of evolution and the experience of practical farmers and ranchers who have little sympathy for unproductive paperwork. If you find the description daunting, rest assured that it is more straightforward to do than to read about. It will work as well for small operations as large ones and brings the greatest return in times of crisis.

Mental attitude is far more important than financial resources. You've got to carry through, be creative but tough-minded, and not fudge to make things look rosier than you know they are. For starters, however, don't think of the task as a drudgery of numbers and balances. Think of yourself as a conductor about to direct a symphony. You are about to exercise that kind of mastery over your own life and a big stretch of land. Art doesn't get more spellbinding than that.

If you can acquire computer programs to handle the main planning sheets, do it. Any sophisticated spreadsheet program will serve. That eliminates the toil of balancing columns and erasing yards of numbers every time you change a detail of the plan. Tedious hours of calculations, locating errors, and redoing work can be very discouraging and will drain your creativity.

Forms

You will need only four kinds of forms:

1. *Worksheets:* Discussed in the previous section, at least one will be used to detail every income and expense item.

2. *Livestock Production Plan:* Discussed in the previous section, this is really a glorified worksheet for detailing livestock breeding operations.

3. *Annual Income and Expense Plan* (planning sheet): This multicolumned spreadsheet relates time, income, expenses, cash, bulk purchase consumable items, and credit. Printed versions are useful for assembling data, but computer spreadsheets can be easily set up to handle the endless modifications and trial plans called for in the planning process.

4. *Control Sheets:* These forms are for reporting items that run counter to plan and for recording decisions on corrective action.

Any other forms or scraps of paper should be avoided. They add to confusion and stress, especially in the most difficult part of planning—control.

Who Should Plan and When

Don't try to plan alone. Recruit a team that is as inclusive as possible without being unwieldy. There are three reasons for this strategy:

- People who become involved with projects at the beginning will care about them and see them through to the end.

- When hard choices must be made, morale will survive much better if everybody understands why and has a chance to work on solutions.

- It takes a lot of work to prepare for good decisions. Budgets, inventories, gross margin analyses, time sequences—everything that requires a worksheet—might keep one person tied up year round. A team, especially if it includes people intimately acquainted with field realities, gets through that work faster and better than any individual.

If you haven't ever done it before, get organized to start planning as soon as possible. After that you routinely start the process about 2 months before the financial year begins.

Planning will take 4 to 8 days of uninterrupted time. It does not have to be (and often can't be) in one block, but allowing too much time between planning sessions takes a toll on momentum. The schedule should be set in advance, so commitments and deadlines mean something. The schedule should include:

- A social event, potluck or barbecue, to establish a team spirit and do the initial brainstorming on such major points as what enterprises to pursue and cost-cutting strategies. Keep the atmosphere light. Include families and children. Don't condemn the ideas that arise. Introduce the process and the schedule to anyone new to the work.

- Time to gather information before the formal planning begins. Alert team members to research everything potentially relevant—sales trends, equipment costs, the opinions of outside experts, inventories, whatever. It can be extremely frustrating to gather for a weekend of planning on an isolated ranch, only to discover that no one has a clue how much a mile of fence or a new tractor will cost.

- A day or two to plan income sources. This could involve breaking for a day or two for additional research.

- A day or two to plan administration and expenses.

- A period for you or your designated helper to record the cash and credit demands implied by your production and expenses.

- Another team meeting to make the adjustments needed to make the plan work. Try to get a computer that can instantaneously reckon the implications of change.

- Time to organize monitoring and control routines.

Planning the Income

Plan your sources of income every year. Prices, markets, weather, and other forces may undermine the strategy that worked last year, but opportunities continually change. If you can harness and focus some good minds on ways to create wealth, you need never think of yourself as a victim of circumstance.

Things to Keep in Mind

Most wealth comes from the ecosystem, but the forms of that wealth are nearly infinite. Don't forget the value of wild animals and plants or aesthetic production such as recreation, photo safaris, urban people's desire to share in farm life, and diversity of crops.

Land is your fundamental resource. Even if your production goal is wilderness, think through the uses and nonuses of every bit of it. Importing grain for a 1-acre chicken factory to pay the mortgage on a 1,000-acre farm makes little sense if 999 acres of cropland lie fallow.

Your production goals shouldn't specify any "tools" other than living organisms. "Profit from livestock" or "profit from field crops" are suitable production goals. But, "profit from winter wheat" is just as short-sighted a goal as "profit from stud mules." Too inflexible.

Don't forget that a landscape description is a critical part of your goal. You can't make long-term progress toward the other two parts if you don't take steps to produce and maintain the landscape.

Human creativity ranks above all other "tools" for maximizing income.

Any source of income that adversely affects goal attainment is unacceptable in the long term, although some cases will require transition in phases.

Step by Step

The following steps will lead to an income plan, expressed in hard numbers, that makes efficient use of resources and people.

Step One:
Determine the planning schedule and inform the people involved.

Step Two:
Organize a brainstorming session at an informal opening event that includes as many people as possible. Then have the planning team cull the list of income ideas (often more than 100) for the dozen or so that fit the team's values, inclinations, and common sense. Usually the best of your present enterprises will be on the short list automatically (unless you have an obvious reason to drop one).

Step Three:
Assess the short list according to the seven testing guidelines of the HRM model. This may take a while, especially the gross margin analysis, but some of the research can be assigned as "homework" and presented as a report. Be careful not to rely solely on the gross margin analysis. Often an enterprise which tops this test fails several of the other tests and should be dropped. Occasionally

an existing enterprise fails a number of tests as well. That should warn you to find a way to get out of it a soon as possible. High-input farming practices are frequent examples of this predicament.

Step Four:
Fill out separate worksheets for each enterprise under consideration, estimating the income and when it would occur. If convenient you might also compute expenses here. This is especially wise if later expense cuts may affect your income (say, cutting fertilizer). At this point in planning, however, you need the gross income figure, not the net figure. For livestock breeding, use the Livestock Production Plan. The previous section has examples and directions.

Step Five:
After you've planned all the enterprises, complete a worksheet for miscellaneous income.

Step Six:
Set up a column of your planning sheet or computer spreadsheet for each enterprise (as you defined enterprises for your gross margin analysis). The arrangement of columns bears some thought. If you intend to compute net managerial income, you might group those that fall under one person's supervision together. Other ways to categorize them will no doubt occur to you.

Step Seven:
Carry the income figures from all of the worksheets to the appropriate columns and months of the planning sheets (or computer spreadsheet), and record them in the plan row. Do it in pencil! A lot will change. If you're using a computer, make hand entries as well until you know the program has no bugs and you have confidence in it.

Step Eight:
Add the planned figures in each column and across each plan row. They must match in the final total income figure planned for the year.

ANNUAL INCOME AND EXPENSE PLAN

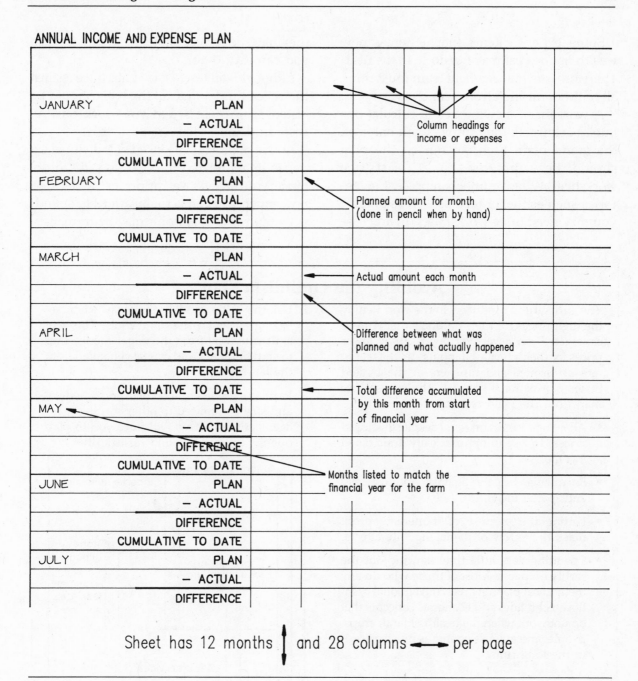

Sheet has 12 months ↕ and 28 columns ◄─► per page

Planning the Expenses

Planning expenses presents an even greater challenge perhaps than planning income. Decisions involving limits and trade-offs always take courage and discipline, but the process outlined here will help keep you and your planning team on target.

Things to Keep in Mind

The expense items include the main agents for generating wealth. Few people realize this, believing that wealth comes only from items marketed. An investment in training, for example, generates wealth but is not itself marketed.

Some expense items that generate new wealth in one year may not do so in the next. Therefore, the management team must think all of them through from scratch every year.

A worksheet showing actual calculations must back up all the numbers you enter on the grand plan. Each column on the plan should be supported by at least one worksheet. Where numbers represent inventories such as fuel or feed as well as dollars, you may need more worksheets. In any case they have to be accessible, neat, and filed so you can find them.

Although you tested each enterprise against the seven guidelines, all the tools should be tested again here when you work out the costs in detail.

At this point you've looked at individual expense items related to what you *like* to do. Since you haven't set priorities or decided how much you can actually afford to spend, your figures at this point may be unrealisti-

---

Avoiding the Optimism Trap

Few agricultural producers in the world enjoy the high prices for their products that Americans get. Even fewer worldwide enjoy lower prices for their purchased inputs. Yet the failure rate of farmers and ranchers in the United States is a national scandal. The excuses are many, but the main reasons are these:

- Allowing production *costs* (despite low input prices) to rise to optimistically anticipated income.

- Borrowing heavily against optimistically anticipated income.

- Letting the promise of immediate profit mask the surety of damaging side effects.

- Spending very little time figuring out the reality on paper. Most of those who do any methodical planning use conventional cash-flow techniques and economic concepts that do not work when the health of land, crops, people, and animals counts as much as cash in the long run.

To counteract these normal habits of mind, you should:

1. Reduce the gross total on your income planning sheet by 50%. Set that as your limit for the living and production expenses you are about to plan. This will discipline you to keep expenses below what you're actually likely to make.

2. Use the HRM model religiously when selecting all enterprises and tools. This will shift you away from the paper money trap toward flexibility and capital generated from the land itself.

3. Work to a predetermined plan and schedule, and stay committed to others on the management team. This will force you to plan openly, thoroughly, and realistically.

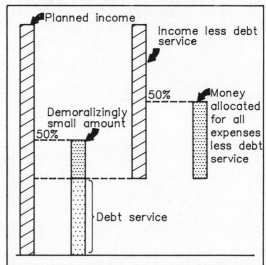

If you carry a big debt, the amount of money remaining after you halve projected income and subtract loan payments may be too small to justify any sort of planning. Since the point of halving income is not to discourage utterly but to encourage thrifty thinking, subtract the debt payment first and halve what's left.

cally high and you may have many more items on the table than you can handle.

The challenge from here on is to cut, paste, compromise, and fit everything into one holistic plan. In doing this you will inevitably modify much of this preparatory work, so don't take it personally when erasers get worn out on things you put a lot of time into. The work was not wasted. Let's say you figured out how to build 30 miles of fence, but end up putting only 20 on the plan. Even so, you can't make that kind of decision at all without thinking through the bigger plan.

Step by Step

Here is a sequence for smelting your best thinking into numbers that will guide the exercise and monitoring of the plan.

Step One:
Decide column headings for the planning sheet. From gathering expense data on worksheets, you'll already know pretty well what your columns will be. But remember: No matter where a particular expense column might fall, you'll want to keep track of all expenses allied to each enterprise for the purpose of gross margin analysis.

The order of expenses on the planning sheet, however, can strongly influence your priorities in allocating funds and analyzing the plan as a whole. Order them according to how directly they contribute to the generation of wealth as discussed in the previous section. You might also arrange expense columns to reflect accounting for net managerial income as described earlier. All expenses under control of one individual or team would go in adjacent columns. For planning purposes the first idea — organizing by direct contribution to generating wealth — probably works best. (Tagging column heads with names, codes, or color marks to indicate NMI categories quite well.)

However you decide to organize your columns, you must decide which items directly generate wealth and which don't. Usually the whole planning team takes part in these decisions. One way is to put the headings on cards and sort them into three piles:

A. Items that would actually generate new wealth (items that cause more solar energy to be converted to dollars). Some items such as training would fall in this category in any year. Others such as fencing or cropland drainage would qualify only if energy conversion were the weak link.

B. Inescapable expenses (fixed amounts that have to be paid—taxes, debt service, and so on).

C. Maintenance expenses (salaries, transportation, supplementary feed, fertilizers, office expenses). Though essential, they do not generate new wealth.

Always plan from a fresh start each year. Something that falls in the A group this year may end up in group C next year or find no place in your plan at all.

Step Two:
Set priorities and allocate funds for the A category. The energy chain from sunlight to solar dollar will have a weak link. The textbook explains this matter in detail and the previous section recaps it again. Some of the expense items in category A will strengthen the weak link in the solar chain. At this point, further testing against the Marginal Reaction per Dollar guideline will commonly eliminate all but one item affecting the solar chain. The rest of the A group, however desirable, will then drop out of the plan completely in that year. (Do not move the A group down to the C category.)

If you expect to strengthen one link quickly, whatever would next become the weak link indicates what you can add to the A group's funding priorities. You might correct a weak marketing link over a short period, for example, leaving the rest of the year open for other action.

Next rank all bona fide A items 1, 2, 3, and so on. Then allocate money to each of them. In doing this you'll have to run everything through the seven testing guidelines yet again. The allocation must pass the Marginal Reaction test and the Energy/Wealth Source and Use test at the same time as the tool on which it is to be spent is itself tested. Let's say you plan to spend money on a bulldozer to clear brush. In this case you must decide whether the practice itself is sound in terms of all tests and whether any money devoted to it will take you to your goals most efficiently.

For those items that pass, try to allocate everything necessary. For the big cost items, allocate all you can, though you may have to adjust later to cover the C items sufficiently. By having gone to the trouble of analyzing your priorities and ranking them, you'll benefit from the natural tendency to spend generously while you have money to spend. By the time you come to the C items, your mood will ensure frugality when you need it.

Step Three:
Allocate what you must to the B items. As these are by definition unavoidable expenses, you don't need to agonize over ranking them. Nevertheless, take the time to assure yourself that they are in fact unavoidable. Financing and structuring debt, for example, is a complex art. Have you really researched and explored all options? Converting debt to equity, equity financing, or lease-back arrangements will not necessarily be suggested by your major creditors. If stuck for ideas, don't hesitate to gather some friends to brainstorm your way out.

Step Four:
Rank the C items and make allocations. Though handled last, these items are impor-

tant and will require funds or your operation will cease to function. But use your imagination to reduce the amount. Since these items do not actually generate new wealth in this planning year, everything you save here is money in the bank. (Some items in last year's

ANNUAL INCOME AND EXPENSE PLAN

	GRASS HAY	ALFALFA USED	BULK GAS	BULK GAS ($)	OTHER GAS
JANUARY PLAN	367 TN		47 G.	$400	$22
– ACTUAL					
DIFFERENCE					
CUMULATIVE DIFFERENCE TO DATE					
FEBRUARY PLAN	132 TN	200 TN	50 G.		$24
– ACTUAL					
DIFFERENCE					
CUMULATIVE DIFFERENCE TO DATE					
MARCH PLAN	80 TN	280 TN	44 G.		$21
– ACTUAL					
DIFFERENCE					
CUMULATIVE DIFFERENCE TO DATE					
APRIL PLAN			49 G.		$23
– ACTUAL					
DIFFERENCE					
CUMULATIVE DIFFERENCE TO DATE					
MAY PLAN			42 G.		$20
– ACTUAL					
DIFFERENCE					
CUMULATIVE DIFFERENCE TO DATE					
JUNE PLAN			42 G.		$20
– ACTUAL					
DIFFERENCE					
CUMULATIVE DIFFERENCE TO DATE					
JULY PLAN			42 G.		$20
– ACTUAL					
DIFFERENCE					
CUMULATIVE DIFFERENCE TO DATE					
AUGUST PLAN			53 G.		$25
– ACTUAL					
DIFFERENCE					
CUMULATIVE DIFFERENCE TO DATE					
SEPTEMBER PLAN			53 G.	$320	$25
– ACTUAL					
DIFFERENCE					
CUMULATIVE DIFFERENCE TO DATE					
OCTOBER PLAN			61 G.		$29
– ACTUAL					
DIFFERENCE					
CUMULATIVE DIFFERENCE TO DATE					
NOVEMBER PLAN	384 TN		42 G.		$20
– ACTUAL					
DIFFERENCE					
CUMULATIVE DIFFERENCE TO DATE					
DECEMBER PLAN	396 TN		42 G.		$20
– ACTUAL					
DIFFERENCE					
CUMULATIVE DIFFERENCE TO DATE					
TOTALS	1,359 TN	480 TN	567 G.	$720	$269

The data from the two worksheets on the facing page would look like this when entered on the annual expense plan. Note figures in both dollars and quantities. The hay and alfalfa worksheet will also contribute to the income plan.

WORKSHEET

CENTER FOR HOLISTIC RESOURCE MANAGEMENT   Date _____ Ranch _____   WORK SHEET NO. _____

Detail	January	February	March	April	May	June	July	August	September	October	November	December	Total
	←FEED AND CALVES→ BULLING→								‹ WEAN	SALE ›	←FEED→		
CATTLE DAYS FED	36,700	33,200	36,800								36,280	39,500	
FEEDING RATE	20LB/DAY	25LB/DAY	30LB/DAY								20LB/DAY	20LB/DAY	
GRASS HAY (TONS)	367	132	80								384	396	1,359
ALFALFA (TONS)		200	280										480
ALFALFA SALE (TONS)		120				480	240					120	960
HAY CUT (TONS)						700			700				1,400
ALFALFA CUT (TONS)						480	480	480					1,440
ALFALFA INCOME		$100/TON $12,000				$80/TON $38,400	$80/TON $19,200					$90/TON $10,200	
Total													

Hay and alfalfa production and consumption.

WORKSHEET

CENTER FOR HOLISTIC RESOURCE MANAGEMENT   Date _____ Ranch _____   WORK SHEET NO. _____

Detail	January	February	March	April	May	June	July	August	September	October	November	December	Total
	←FEED AND CALVES→ BULLING→								‹ WEAN	SALE ›	←FEED→		
MANAGER JEEP	200	250	250	250	250	250	250	250	250	400	100	100	2,800
HERRERA PICKUP MI.	200	200	150	50	50	50	50	50	50	50	200	200	1,300
FOURNIER PICKUP MI.	350	350	250	250	60	60	60	60	60	60	350	350	2,260
SMITH PICKUP MI.	100	100	100	300	300	300	300	500	500	500	100	100	3,200
TOTAL MILES	850	900	750	850	660	660	660	860	860	1010	750	750	9,560 MI.
GALS. GAS @ 12 MI./GAL.	71 G.	75 G.	63 G.	71 G.	55 G.	55 G.	55 G.	72 G.	72 G.	84 G.	63 G.	63 G.	797 G.
HERRERA MI. BIKE GALS.	—	—	100 1.2 G.	100 1.2 G.	300 3.6 G.	300 3.6 G.	300 3.6 G.	300 3.6 G.	300 3.6 G.	300 3.6 G.	—	—	
FOURNIER MI. BIKE GALS.	—	—	150 1.8 G.	150 1.8 G.	400 4.8 G.	400 4.8 G.	400 4.8 G.	400 4.8 G.	400 4.8 G.	400 4.8 G.	—	—	
TOTAL GALS.	71 G.	75 G.	66 G.	74 G.	63 G.	63 G.	63 G.	80 G.	80 G.	92 G.	63 G.	63 G.	853 G.
1/3 OFF RANCH @ 95¢/GAL.	$22	$24	$21	$23	$20	$20	$20	$25	$25	$29	$20	$20	$269
BULK GAS @ 80¢/GAL.	500 G. $400								400 G. $320				
RANCH USE GALLONS	47 G.	50 G.	44 G.	49 G.	42 G.	42 G.	42 G.	53 G.	53 G.	61 G.	42 G.	42 G.	
Total													

Gasoline use and purchase projections.

A group may not even appear in your columns this year, as they're not even maintenance items now. Fence building is a common example.)

For all the C items use the cost-cutting ideas from the brainstorming session that presumably took place at the beginning of the planning process. Of the dozens of ideas that came out of that session, you should have developed the best ones enough to apply them now.

Arrange all the C items in order of priority and set up columns on the planning sheet. For items such as fuel that may involve a dollar amount for the bulk purchases and a monthly consumption in gallons, you may need two columns.

Make allocations up to your self-imposed limit. Toward the end of this process, you may have to juggle between columns to spread the dregs of your resources across all items needing attention. It may even be necessary to go back to the A group and make adjustments.

Remember two warnings:

- Although you've deliberately organized your planning to keep production costs from rising to anticipated income, your suppliers, salespeople, and consultants have not. They will still push strategies for "maximum gain," "heavy calves," "higher yield," and so forth instead of "greater profit." You can use their expertise, but make everything clear at the start: "Here are my limits. What can you do within them?"

- All figures on the planning sheet must be worked out on a worksheet, and these worksheets, containing all your reckoning, must be available for later control.

Step Five:
Transfer the monthly figures from the worksheets to the plan. Up to now you've only dealt with annual totals. Now you will fill in the monthly amounts in the "plan" rows on the master planning sheet, including both dollar and quantity figures.

Step Six:
Depreciation. After planning all columns, add one more (backed by a worksheet) and allocate money for the replacement of machinery. This step is optional, but it's wise to treat it as an actual expense and make payments into an interest-bearing account. It assures that you won't have to borrow when things wear out. Work it out according to what you know about the life of your machines, not tax amortization schedules.

Step Seven:
Check your figures. Total the columns and rows. The sum of the row totals and the sum of the column totals must be the same. Ferret out the mistakes and enter the grand total in the lower right corner of the planning sheet. If you do this by calculator, use one that prints a tape—it makes finding errors vastly easier.

Managing the Cash

You have now laid out your proposed business year in its entirety. Take a good look at it. Even if the bottom line comes out positive, you still do not know if you will have cash on hand when you need it or what your credit needs might be.

Do your cash management analysis either to the right of all expense columns or on the planning sheet with the income columns, which usually has more free space. If you handle financing through a simple overdraft arrangement with your bank, you'll need three columns. If you handle financing through notes, you'll need five. If you use both kinds of credit, you'll need six. Frequently an operation uses several sources of finance involving notes, bank lines of credit, and co-op credit source. You can set up separate columns for all of these, but that tends to obscure the total picture you need for plan-

ANNUAL INCOME AND EXPENSE PLAN

	SURPLUS (DEFICIT)	BANK BAL. $250,000	BANK INT. OWED 15%	NOTE ↑↓	NOTE INT. 12%	NOTE BAL. 300,000
JANUARY PLAN	(35,000)	175,000		40,000	3,000	263,000
– ACTUAL						
DIFFERENCE						
CUMULATIVE DIFFERENCE TO DATE						
FEBRUARY PLAN	(20,000)	155,000			2,630	265,630
– ACTUAL						
DIFFERENCE						
CUMULATIVE DIFFERENCE TO DATE						
MARCH PLAN	(35,000)	120,000			2,656	268,286
– ACTUAL						
DIFFERENCE						
CUMULATIVE DIFFERENCE TO DATE						
APRIL PLAN	(35,000)	85,000			2,683	270,969
– ACTUAL						
DIFFERENCE						
CUMULATIVE DIFFERENCE TO DATE						
MAY PLAN	(30,000)	55,000			2,710	273,679
– ACTUAL						
DIFFERENCE						
CUMULATIVE DIFFERENCE TO DATE						
JUNE PLAN	5,000	60,000			2,737	276,416
– ACTUAL						
DIFFERENCE						
CUMULATIVE DIFFERENCE TO DATE						
JULY PLAN	1,000	21,000		40,000	2,764	239,180
– ACTUAL						
DIFFERENCE						
CUMULATIVE DIFFERENCE TO DATE						
AUGUST PLAN	(20,000)	1,000			2,392	241,572
– ACTUAL						
DIFFERENCE						
CUMULATIVE DIFFERENCE TO DATE						
SEPTEMBER PLAN	(20,000)	(19,000)	238		2,416	243,988
– ACTUAL						
DIFFERENCE						
CUMULATIVE DIFFERENCE TO DATE						
OCTOBER PLAN	(20,000)	(39,238)*	490		2,440	246,428
– ACTUAL						
DIFFERENCE						
CUMULATIVE DIFFERENCE TO DATE						
NOVEMBER PLAN	260,000	220,272			2,464	248,892
– ACTUAL						
DIFFERENCE						
CUMULATIVE DIFFERENCE TO DATE						
DECEMBER PLAN	(15,000)	205,272			2,489	251,381
– ACTUAL						
DIFFERENCE						
CUMULATIVE DIFFERENCE TO DATE						
TOTAL						

Cash management plan form.

*When the bank balance is negative, as in September ($19,000), remember to add the interest owed ($238) to your debt. If the month's business shows a loss, as in October ($20,000), add the loss to your bank debt as well. Example: $19,000 + $238 + $20,000 = $39,238.

ning. Try to combine overdraft arrangements and notes under two composite headings, using average interest rates if necessary.

Headings for overdraft arrangements are: Monthly Surplus/Deficit, Bank Balance, and Interest Owed.

For notes in combination with an overdraft arrangement, you need six columns: Monthly Surplus/Deficit, Bank Balance, Interest on Overdraft, Note Increase/Decrease, Note Interest, and Note Balance.

For each month, total your income and expense figures in pencil on the "planned" line. Then proceed as follows:

1. From the production and expense sheets, compute your planned *cash* surplus/deficit each month. *Don't add month to month.*

2. Enter your beginning bank balance above column 2 ($250,000 in example).

3. Enter your scheduled payments on notes in column 4 (two payments of $40,000 in example).

4. Enter your beginning note balance above column 6 ([$300,000] in example).

5. Compute your monthly cash position as follows:

   a. Figure the note interest using one-twelfth the annual rate and enter in column 5 (for example, 15% annual/12 months = 1.25% monthly).

   b. Add interest and *subtract* payments to find note balance for next month.

   c. Subtract your note payments from the bank balance, and add your monthly cash surpluses or subtract your deficits.

   d. If your bank balance goes into overdraft, use one-twelfth the annual interest rate to find the interest. Enter in the month of overdraft.

   e. If a month showing a *positive* bank balance follows a month showing a *negative* bank balance, remember to reduce the positive balance by the amount of interest owed from the previous month.

The numbers shown here represent operating expenses financed by an overdraft arrangement at a local bank alongside a long-term note. The "note" could in fact be a combination of Land Bank loans or equipment loans. For simplicity's sake they are combined here into one column with two predictable-as-death $40,000 payments.

The operator has advanced his cash position about $4,000. Could he do better by managing his cash differently? Given the cash flow generated by production, could this debt be handled more economically? Would a different weaning and marketing strategy produce a more consistent cash flow? To what purpose? Could short-term cash surpluses be put to work in other ways?

Analyzing the Plan

Apart from the cash management prediction, you have to make sure that the production policy is sound from an overall business point of view. Will it produce a profit or loss in real terms at the end of the planned period? Will it leave you in an acceptable position in terms of your goals?

To answer these questions, you must produce (with an accountant's help if necessary) an "estimated profit and loss statement" to account for noncash changes in your financial position. An increase in livestock inventory amounts to an increase. Reductions in livestock and depreciation of equipment (if not already included on the planning sheet) are decreases.

Again this statement will not be quite the same as the one you'll use for tax purposes or a loan application, since it doesn't include such noncash items as changes in the value of real estate assets, mineral and water rights, and securities. Although many enterprises depend on such gains, they don't represent "solar wealth," may prove fickle, and often

aren't liquid enough to afford help if you need it. Even if you use these paper assets to underwrite capital improvements that will boost "solar productivity," you're treading on thin ice.

If the year's plan looks good, then proceed. If not, then replan right away. Replan until the results satisfy you. This replanning requires a great many adjustments and erasures of all those figures you wrote in pencil on the planning sheets (if you do everything by hand). Now, just when you need all the energy and creativity you can muster, you can easily fall into bitter arguments over how to cut and paste. Before that happens, consider: Many planning teams have found that extremely well-articulated arguments often have nothing to do with the plan. They are trumped up mainly to avoid the agony of reworking the plan. If you haven't already done so, consider getting a computer. Then you only have to fight over who gets to use the machine. Just punch in the suggestions, and see if they work.

Control Through the Year

No plan ever goes exactly to plan. No planning is complete without monitoring, controlling, and replanning. Because of these two points, proceed as follows:

1. Establish the most convenient means to obtain, before the tenth of each month, the actual figures of income, expenses, and consumption of inventory items.

2. Enter your figures *in ink* on the second row (headed "Actual").

3. For each month compute the difference between the planned and the actual

ANNUAL INCOME PLAN

		COMMERC. CATTLE	ALFALFA HARVEST	ALFALFA INCOME
OCTOBER	PLAN	$9,000		
	− ACTUAL	9,750		
	DIFFERENCE	750		
CUMULATIVE DIFFERENCE TO DATE		750		
NOVEMBER	PLAN	325,156		
	− ACTUAL	321,025		
	DIFFERENCE	<4,131>		
CUMULATIVE DIFFERENCE TO DATE		≪3,381≫		
DECEMBER	PLAN			$10,800
	− ACTUAL			9,600
	DIFFERENCE			<1,200>
CUMULATIVE DIFFERENCE TO DATE				≪1,200≫
JANUARY	PLAN			
	− ACTUAL			
	DIFFERENCE			
CUMULATIVE DIFFERENCE TO DATE				
FEBRUARY	PLAN			$12,000
	− ACTUAL			14,400
	DIFFERENCE			2,400
CUMULATIVE DIFFERENCE TO DATE				1,200
MARCH	PLAN			
	− ACTUAL			
	DIFFERENCE			
CUMULATIVE DIFFERENCE TO DATE				
APRIL	PLAN			
	− ACTUAL			
	DIFFERENCE			
CUMULATIVE DIFFERENCE TO DATE				
MAY	PLAN			
	− ACTUAL			
	DIFFERENCE			
CUMULATIVE DIFFERENCE TO DATE				
JUNE	PLAN		480 TN	$38,400
	− ACTUAL		400	32,000
	DIFFERENCE		<80>	<6,400>
CUMULATIVE DIFFERENCE TO DATE			≪80≫	≪5,200≫
JULY	PLAN		480	$19,200
	− ACTUAL		500	23,400
	DIFFERENCE		20	4,200
CUMULATIVE DIFFERENCE TO DATE			≪60≫	≪1,000≫
AUGUST	PLAN		480	
	− ACTUAL		510	
	DIFFERENCE		30	
CUMULATIVE DIFFERENCE TO DATE			≪30≫	
SEPTEMBER	PLAN			
	− ACTUAL			
	PLANNED DIFFERENCE			
CUMULATIVE DIFFERENCE TO DATE				
(PLANNED)	TOTAL	$334,156	1,440 TN	$80,400

Plan shows relationship between planned and actual figures for cattle and alfalfa.

4. After the first month record the accumulated difference in row 4, again using red or blue ink. This is the sum of the differences in previous months. It will alert you when small differences ignored in your monthly control represent a serious drift away from your plan.

This is *not* an accounting procedure. Rough figures *now* serve better than perfection a month late. You may use the normal bookkeeping process, but sometimes check stubs or other tallies work well enough and faster.

One look at the planning sheet and its red and blue entries will show you the deviations from plan at a glance. Consider all the major adverse deviations in detail, column by column, going back to the original paperwork (which you have carefully filed) as necessary.

At first the deviations may derive from inexperience, but as the years pass you'll get better. In any case . . . do something now!

If the *income* items are seriously adverse, they can only be "controlled" by cutting total expenses. This will require you to replan, perhaps from scratch. If the *expense* items are adverse, always control the item itself. You must never simply eyeball the statistics and balance a surplus in one column against a deficit in another. If you don't keep every expense on target, you'll soon lose control of the whole plan.

It is crucial to apply the utmost energy and imagination to get things back on track. Mental attitude counts more than anything. Don't hurry.

amounts. Enter the result *in ink* in the third row (headed "Difference"). For all income and expense columns, enter the figures that are adverse to plan in red ink and those better than plan in blue so your overall position will be graphic.

Don't panic. Think, think, think. Plan, plan, plan. Neither make, nor accept, any excuses. The word "can't" must not enter your head.

Serious deviations from plan also constitute the greatest danger to management relations and the morale of your whole staff—another reason for careful monitoring. The temptation to assign blame quickly and chew somebody out will be great, but likely as not will assure your ultimate failure.

The best solutions often lie in a flexible and cooperative sharing of time and resources commanded by several staff members rather than in merely turning up the heat on the one person in whose bailiwick the problem technically occurs. Anything you do now to exacerbate unhealthy competition, resentment, or turf battles among staff could prove fatal to your enterprise.

Make the problem a matter of shared concern and responsibility, and invite a team approach to solving it. This is where the "goal ownership" you built in the early planning stages pays off in spades. Every crisis offers an opportunity for staff members to show their stuff. And seeking their input, rewarding creativity, and delegating authority builds morale as surely as tyranny tears it down.

Generally—and certainly where a management team answers to a board or absentee owner—a monthly Control Sheet should be completed. This form states the column heading or number, the amount of deviation to date (in dollars, gallons, tons, and so forth), the cause of the deviation, and the proposed action.

Last, but most important, decide who is to act and record this person's initials in the action column.

Center for Holistic Resource Management

CONTROL SHEET

Name: _____ Date: *June 1988* Sheet #: *1.*

Plan Column #	Amount Adverse to Date	Cause of Deviation from Plan	Proposed Action to Return to Plan	ACT
I-5	$5000	Sales prices lower than expected.	- Plan reductions in all areas — each person to prepare & come up with suggested reductions by next control meeting.	All
			- Sell as pairs next year & market wider afield.	B.J.
E-16	$1500	Labor used earlier than planned.	- No action - will balance in August.	-
E-26	360 gallons	Poor use of heavy transport. Too many unnecessary trips.	- Reduce to 1 trip to town each month. Keep purchase requests at office.	B.J.
E-37	$636	Farrier charges have risen as has transport cost.	- Horses will not be shod. All hooves will be rasped & checked weekly. Send Jack on course in November.	J.S.
E-39	200 gals. gas	Too much pickup use.	- Supplies will be bulked in small storage sheds at strategic sites on ranch.	
			- Use 4-wheelers - no pickup use in good weather except emergency.	M.K.
E-42	$2000	Several unexpected breakdowns with machinery, pickups & tractor.	- Arrange brainstorming party Tuesday next week at Bill's house 6 p.m. All staff present.	B.J.

SUMMARY

Earlier I remarked that money itself is neither wealth nor happiness nor the fulfillment of dreams, but it may be a measure of progress along the road to some of these things, and so you must plan. When farms and ranches representing the lives and dreams of generations go down to the mournful echo of the auctioneer's hammer because no one saw the danger ahead, we talk of economics as an evil that grabs people by the throat in the dark. Yet you and I know that is not true. Just turn on the lights. Do the financial planning. The monster vanishes.

You can't manage in the dark. If you can't see all the pieces at once, even the notion of holism becomes absurd. Do the financial planning.

Allan Savory, who originated these procedures and has seen them return many an apparently hopeless case to the glory road of prosperity, often rails against a lack of self-discipline of well-endowed managers who don't plan. The people whose bones now bleach beside the road (more men than women) spent too much time drinking coffee and tinkering with machinery. Do the financial planning.

For my part I don't think most nonplanners lack self-discipline. When I don't plan, it is often because between me and those I care for, be they family or associates, there is a no man's land of competition, disagreement, or ambiquity that we dare not enter. Nothing lights a fuse quicker than money.

"If my wife knew the size of our debt, she'd freak out."

"Maybe, if we have enough left after the calves are sold, I'll get my daughter a car to take to college, or we can go visit my wife's parents, or I'll lay that pipe from the spring."

If you wait till fall and the money is or isn't there, then you won't have to argue about it. Right? Or if there is an argument, the winners will spend the money and what happens then will be *their* fault. Either way, why stir things up now?

Because if you do, and you really can bring people together, set goals, and eliminate the no man's land of silence, the planning won't require any self-discipline at all. Economics, which is no less a part of life than love, will be on your side. Switch on the lights and do the financial planning.

PART II
BIOLOGICAL PLANNING

BIOLOGICAL PLANNING

This part tells about biological planning, not land planning. Though both aspects belong to one process, along with financial planning, confusing the two can lead to problems.

Not so long ago people wishing to practice holistic management using livestock were usually advised to build an arrangement of fences in the pattern of a wagon wheel around a central water point. Although this is indeed an efficient way to handle livestock in many situations, the wagon wheel became such a strong symbol that it obscured the idea behind it. Like people who draw witch signs on the doorpost to keep away lightning, many ranchers and researchers built wagon-wheel grazing cells in the belief that they had magic power to grow grass.

A legacy of the wagon-wheel dogma is the term "cell," which Allan Savory first applied because one of his early projects looked like the top view of a honeycomb. As the practice of holistic management has evolved, however, this connection has been lost. Terrain, aesthetic and cultural goals, finances, water availability, and other factors often make fences, not to mention the wagon-wheel pattern, unwise. And even where a particular form does make sense, it should never function as an isolated entity outside larger "wholes."

Planning for livestock grazing, not fencing, defines a cell. In holistic management, a cell is any piece of land planned as a unit. It is the medium through which your goals will be sought and realized, and the management of a cell requires consideration of all the factors that influence progress toward those goals. Nevertheless, no cell constitutes a complete world. As your goals and circumstances evolve, you *will* redefine the cells.

Art is sometimes defined as the representation of truth and beauty within artificial limits—a frame, the pages of a book, the range of an instrument. It is a daunting illustration of "progress" that not long ago whole continents were "cells" in which all interactions occurred without artificial limit, but today we measure cells in mere acres.

Planning for such small units demands considerable art, because such plans cannot be perfect, especially in the initial stages of land development. You may well have to settle for some degree of overgrazing. You may well find it impossible to apply all the tools to maximum effect. This can prove discouraging enough to put some people off planning altogether. But holistic thinking means embracing reality, not waiting for a perfect context. Only the experience of making the best plan for the season ahead can show you how to undertake long-range land planning to push back the limits and give yourself a more generous frame for your art in years to come.

Part IV, "Land Planning," addresses that broader challenge. Today, however, success lies in doing the best you can with what you have.

MASTERING THE BASICS

Carrying out the financial plans outlined in Part I calls for a control of detail beyond anything generally seen in conventional management. The main difference, however, is in the kind of detail you control. In holistic management of livestock and game, the traditional goal of "raising meat" becomes a by-product of more primary purposes—creating a landscape and harvesting sunlight. This section deals with that challenge at the field level. Since the HRM textbook explains the principles in depth, read it carefully. Here is a review of the major operational points and examples of how to apply them.

The HRM model puts all land management tools into six basic categories. Of these six, rest, grazing, and animal impact are the primary subject of biological planning. Four guidelines govern most management decisions concerning these tools—population management (stocking rate), time, stock density, and herd effect. Only the first three guidelines can be quantified. (Herd effect is a matter of behavior.) Division of the cell into grazing areas is essential, too. *Though these areas need not be fenced "paddocks," we'll use that word for simplicity's sake.*

Most planning centers on these questions:

- What is your landscape goal?

- How much total forage will the cell have to supply?

- How much forage will an average acre of land have to supply?

- How rapidly will animals deplete forage on each acre?

- How long will standing forage at the end of the growing season last (including reserves for late springs, drought, fires, wildlife, and so on)?

- How long will animals spend in each paddock, and when will they return?

Measuring Range Use and Forage

Standard American practice measures range use in "animal unit months" (AUM). Months work well for computing grazing fees, but they prove clumsy for reckoning whether a herd should move in 2 days or 3. In this case an "animal unit day"—animal-day (AD), for short—works much better.

An animal-day is the *amount* an animal eats in a day and is thus first of all a measure of forage quantity, not animals or days. In all your figuring, the animal-day will have the same character as bales, pounds, tons, cubic feet, or whatever unit you use for bulk forage.

Unlike a ton or a cubic yard, an animal-day of growing forage has no status at the Bureau

---

Animal-days of Grazing

A cattle-day is the food a dry cow eats in one day.

Animal-days = Animals x Days

A sheep-day is the food a grown sheep eats in one day.
About 5 sheep-days = 1 cattle-day

1 cow x 100 days = 100 animal-days of forage
10 cows x 10 days = 100 animal-days of forage
4 cows x 15 days = 60 animal-days of forage

of Standards. Fortunately animals have a gut feeling for the exact amount. They simply take it from the land every day. A cattle-day, roughly speaking, is enough forage to fill a cow's stomach. Your sense of what that means will grow quickly as you compute how many animal-days your herd has taken and observe the land before and after.

Cattle of course eat more than sheep, so you might have sheep-days (ShD) for sheep and cattle-days (CD) for cattle. For more precision, both can be translated into standard units that reflect the kind of animal, its weight, rate of gain, reproductive status, and so on. This section includes tables for translating the head count into "standard animal units" (SAU) that weigh the demands of each class of stock in relationship to the appetite of a standard cow. Thus we also have the "stock day" (SD) as a standard unit of forage.

If you forgo the work of reducing your herd to standard animals, don't forget that a prize bull or lactating cow takes far more from the range than a young steer. Though the text may not always distinquish between AD and SD, there is a difference.

The concept of animal-days helps you:

- Estimate how much forage stock or wildlife will require from a given section of land

- Plan for dormant season and emergency reserves

- Assess the impact of grazing on areas of different quality (or pressure on animals to graze unpalatable or poisonous plants)

- Determine a realistic stocking rate

Animal-days per Acre (ADA)

Ideally you'd like to know what each plant will have to contribute to the animal-days of forage your herd will take. The "animal-day per acre" (ADA) is a practical alternative.

Herds of different sizes may spend varying lengths of time in paddocks of different sizes, but you can still reckon how much forage the average acre in that paddock will supply.

Animal-days per Acre of Grazing

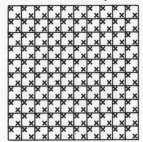

Animal-days per acre = $\dfrac{\text{animals x days}}{\text{acres of land}}$

Example: 50 animals spending 4 days in a 100-acre paddock will take 50 x 4 ÷ 100 = 2 ADA from that paddock. An average acre will yield enough forage to feed 2 animals for a day.

4 days
50 cows on 100 acres

Example: 50 animals spending 4 days in a 50-acre paddock will take 50 x 4 ÷ 50 = 4 ADA. An average acre will feed 4 animals for a day.

50 cows on 50 acres
4 days

The following example shows something of what this means in practice. In a winter's grazing, cattle have been through all the paddocks in a cell three times. In one 500-acre paddock, 100 head spent 3 days, 4 days, and 3 days. The same 100 head spent 1 day, 1 day, and 1 day on a 100-acre piece of bottomland, but on the last day they were joined by an extra 100 head gathered for shipping. Both paddocks had acceptable amounts of litter and forage left after the last grazing. You want to know the relative production from these two paddocks.

Working Out Animal-days/Acre Yield of Land

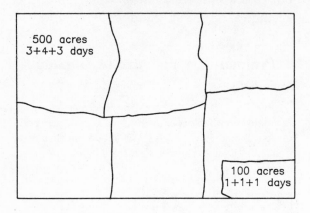

Land in the first yielded:

$$\frac{100 \text{ animals} \times 10 \text{ days}}{500 \text{ acres}} = 2 \text{ ADA}$$

Land in the second yielded:

$$\frac{(100 \text{ animals} \times 3 \text{ days}) + (100 \text{ animals} \times 1 \text{ day})}{100 \text{ acres}}$$

$$= 4 \text{ ADA}$$

Obviously if one paddock seems to suffer more than others (or advances more rapidly in succession), you can compute the ADA taken from it and change your plans.

Conversion into Standard Animal Units (SAU)

For a more useful measure of forage consumption you'll have to convert to standard animals. This becomes particularly important when you're computing the needs of breeding herds when feed is scarce. The accompanying tables give conversion factors for sheep and cattle. Essentially a "standard animal" is a 500-pound steer gaining a pound a day or a 1000-pound dry pregnant cow. Everything else takes either more or less.

The cattle table gives conversion factors for different rates of gain. You cannot eyeball the chart and say, "Oh, I want my steers to gain 1.65 pounds/day." Wishful thinking won't change genes and forage quality. But if experience tells you that, given the forage, they can do that well, use the numbers in that column.

The planning procedure at the end of this part ("Creating Your Plan") calls for extensive use of these tables. Below is an example to show how the conversion is done.

Working Out Standard Animal Units

Detail	Jan	Feb	Mar	Apr	May	Jun	Jul	Aug	Sept	Oct	Nov	Dec
Bio Year	///// CALVES /////			\\\\\ BULLS \\\\\					⟨WEAN			
900# Cows	500	500	500	700	700	700	700	700	700	686	548	548
Phys Factor	1.15	1.30	1.54	1.54	1.54	1.54	1.54	1.30	.95	.95	.95	.95
S.A.U.	575	650	770	1,078	1,078	1,078	1,078	910	665	652	521	521

The physiological factor for an 891-pound dry pregnant cow is 0.95; for a lactating cow the figure is 1.54. Interpolating for the first month of calving, we might say 1.15 x 500 cows = 575 standard animal units.

Standard Animal Unit Tables

PHYSIOLOGICAL FACTOR — CATTLE

Est. Weight		Steer/heifer desired daily weight gain, kgs/lbs				Lactating cow 3–4 months	Pregnant cow	Bull
Kgs.	Lbs.	0.75/1.65	0.5/1.1	0.25/0.55	0/0			
150	330	0.73	0.67	0.57	0.44			
180	396	0.93	0.79	0.67	0.51			
225	495	1.18	0.99	0.81	0.60			
270	594	1.38	1.18	0.94	0.69			
315	693	1.56	1.35	1.05	0.77			1.66
360	792	1.74	1.52	1.17	0.85	1.43	0.88	1.78
405	891					1.54	0.95	1.88
450	990					1.62	0.98	1.97
495	1089					1.70	1.07	2.00
540	1188						1.12	2.00
585	1287						1.21	2.00

PHYSIOLOGICAL FACTOR — SHEEP*

Est. Weight		Ewes In moderate condition. For fat ewes use 10 kgs lower weight						Rams G+	Lambs	
Kgs.	Lbs.	A	B	C	D	E	F+		Early weaned	Fattening
10	22								0.63	
15	33								0.84	
20	44								1.04	
25	55								1.25	
30	66	0.65					1.15		1.46	1.19
35	77	0.69					1.18			1.34
40	88	0.72					1.21	1.68		1.61
45	99	0.75					1.20	1.76		1.71
50	110	0.79	0.86	1.41	1.96	2.23	1.18	1.83		1.81
55	121	0.83	0.94	1.50	2.05	2.33	1.18	1.91		1.91
60	132	0.86	1.02	1.58	2.14	2.42	1.18	1.98		
65	143	0.90	1.06	1.66	2.24	2.52	1.14	2.04		
70	154	0.94	1.10	1.74	2.33	2.61	1.10	2.09		
75	165	0.98	1.14	1.78	2.38	2.71		2.15		
80	176	1.02	1.18	1.82	2.42	2.80		2.20		
85	187			1.87				2.20		
90	198			1.91				2.20		

Legend (SHEEP):

Nonlactating ewes—Maintenance (in moderate condition) ....A
First 15 weeks gestation ....B
Last 6 weeks gestation C

Lactating ewes—Suckling single lambs
First 8 weeks lactation..............................D
Last 8 weeks lactation C

Lactating ewes—Suckling twin lambs
First 8 weeks lactation ....E
Last 8 weeks lactation ....D

Ewes, replacement lambs, and yearlings...........F

*Weight given per head while all SAU figures pertain to 5 head each of that weight.
+Values for replacement lambs (rams and ewes) start at time they are weaned.

—*Abridged from tables compiled by A.H. Penderis*

Grazing and Growth Rates

Simply stated, overgrazing occurs when an animal bites off a plant before it has recovered from the last severe bite. This second bite weakens the plant because it must sacrifice energy stored in its roots to recover from any severe defoliation during growth, and successive losses without recovery can destroy the roots altogether.

> *Overgrazing happens when animals linger too long among rapidly growing plants—or if they return too soon when growth is slow.*

The recovery requirement varies from as little as 10 days on certain irrigated pastures under ideal conditions to 90 days or more on rangeland during dry or cold weather. Using times typical for brittle ranges worldwide, the diagrams show how this principle works.

In practice you will flag severely grazed plants and monitor actual growth rate. If your landscape goal calls for complex, stable grassland, you'll adjust the timing to the most severely grazed perennial grass plant. Severely grazed plants recover much more slowly than moderately grazed ones as the graph below shows. You must judge by the worst case, even when the majority of plants fare better.

A dilemma always arises over grasses that recover at different rates because of species characteristics, slope direction, or other causes. If you have only a few paddocks—and thus long grazing periods—you may have to risk overgrazing some plants. Rapid moves favor the fast growers, but the herd returns too soon for the slow ones. Slow moves may expose fast growers to a second bite. You'll have to choose the best compromise for your livestock and land. (The problem becomes less acute as paddock numbers increase.)

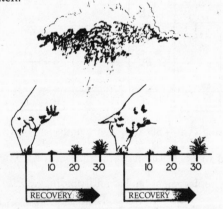

During poor growth conditions on rangeland, plants may need 90 days or more to recover after being severely bitten.

When proper moisture, temperature, and aeration combine, grass may grow very fast and needs less recovery time. Animals may bite it off after as little as 30 days.

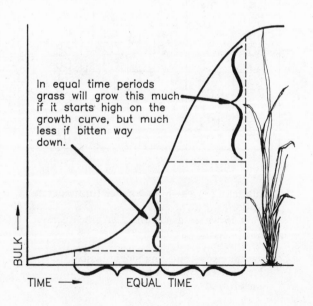

In equal time periods grass will grow this much if it starts high on the growth curve, but much less if bitten way down.

Grazing Periods and Recovery Periods

Overgrazing depends primarily on the recovery period. Once you've decided the recovery period, you can determine the grazing periods in each paddock. In a cell of six equal paddocks, for example, a 90-day recovery period dictates 18-day average grazing periods.

Figuring Grazing Periods

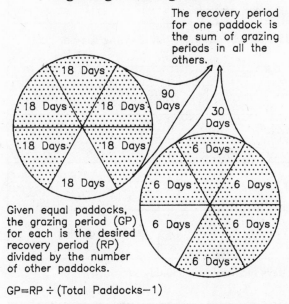

The recovery period for one paddock is the sum of grazing periods in all the others.

90 Days

30 Days

Given equal paddocks, the grazing period (GP) for each is the desired recovery period (RP) divided by the number of other paddocks.

GP=RP ÷ (Total Paddocks−1)

During rapid growth, severely bitten plants might require only a 30-day recovery period. Thus grazing periods would fall to 6 days. The rule of thumb is:

$$\frac{\text{Slow growth} = \text{slow moves}}{\text{Rapid growth} = \text{rapid moves}}$$

When in doubt, use slow moves.

This rule runs counter to instinct. You naturally want to keep animals longer on flushing grass and move on sooner when it's brown—but every time you leave a paddock a day early, it cuts a day off the recovery time *for all paddocks*.

Time, Paddocks, and Land Divisions

Timing and planning—these are the keys to holistic grazing management. As the textbook explains in detail, overgrazing occurs when animals take regrowing foliage before the plant has recovered from a previous bite. This happens when animals linger too long in one area or return to it too soon. Since plants grow at different rates, timing demands thought and vigilance even without such complications as calving and lambing, poisonous plant seasons, water scarcity, weather, and competing land uses.

Nevertheless, most landscape and production goals require you to consider all factors in the primary context of stopping overgrazing. Land divisions are the instrument for doing that. They may be fences, natural barriers, or just a visual demarcation for herders.

Whether you're running a 60-paddock radial cell, leased land transected by ancient barbed wire, or unfenced range, you will have to decide how long your animals will stay together in one place, how big that place will be, where they will move next, and when they will come back. During the growing season, this planning is absolutely critical, but it merits serious reflection in the non-growing season as well.

The Effect of Paddock Numbers on Timing

As the following diagrams show, the more divisions of land you have, the more recovery time per day of grazing each paddock gets. Increasing the number of paddocks doesn't change the ADA yielded by the cell as a whole. It just means that each acre gives up its share in a shorter time followed by a longer recovery period. Increasing the number of paddocks also lessens the danger of overgrazing if recovery periods are kept long during rapid growth.

More Paddocks Increase Available Recovery Time

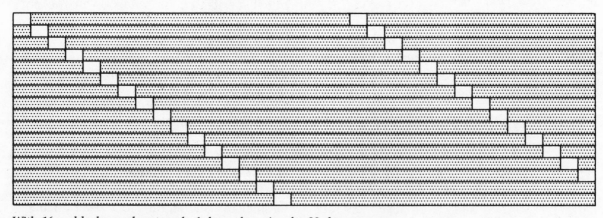

90 Days   180 Days

Paddock 1——►
90 days grazing

90 days recovery

Paddock 2——►
90 days recovery

90 days grazing

With 2 paddocks, each may get 90 days grazing and 90 days recovery in a 180-day growing season.

Paddock 1——►
30 days grazing

90 days recovery

Paddock 2——►

30 days grazing

Paddock 3——►

30 days grazing

Paddock 4——►

30 days grazing

With 4 paddocks, each could get 30 days grazing and 90 days recovery in a 180-day season.

With 16 paddocks, each gets only 6 days of grazing for 90 days recovery.

Effects of Timing with Various Paddock Numbers on a Typical Brittle Range

| Paddocks | Days Grazing | Days Recovery | Rapid Growth | | Slow Growth | |
			Grazing Period	Recovery Period	Grazing Period	Recovery Period
8	4	30	GOOD	GOOD	GOOD	≫DANGER≪ too short
			(Few plants grazed at low stock density.)			
8	13	90	DANGER regrowth regrazed	GOOD	GOOD	GOOD
				(More plants grazed as time increased.)		
31	1	30	GOOD	GOOD	GOOD	≫EXTREME≪ ≫OVERGRAZING≪
				(More plants grazed at higher stock density.)		
31	3	90	FAIR	GOOD	GOOD	GOOD
				(Even more plants grazed at higher stock density and longer time.)		
91	1	90	GOOD	GOOD	GOOD	GOOD
				(Many plants grazed at very high stock density.)		

To extend the recovery period in one paddock you must extend the grazing period in several. In an eight-paddock cell mistakes can hurt. A 90-day recovery period means 13-day average grazing periods—which means that rapidly growing plants bitten severely on the first day may get bitten two or more times on regrowth.

As the number of paddocks increases, the grazing periods become too short for overgrazing to occur, but other dangers increase as the chart above shows. Note that at 30 paddocks or so, slow moves during fast growth cease to cause overgrazing. But at high paddock numbers the damage caused by moving too fast during *slow growth* increases dramatically. Moving too fast is a great temptation because a large number of animals crowded into a small area will deplete it quickly, and moving them a day early leaves a paddock looking much better. However, this cuts the recovery period by a month in a 30-paddock cell, and the high-density herd may overgraze every single plant.

Experience has shown that in terms of solar dollars obtained through livestock, land divisions are usually the most productive investment initially. Up to about 30 paddocks, each new division significantly shortens the average grazing period. After that, the time gains taper off. As we will see, additional paddocks continue to increase stock density, and that

Paddock Number and Grazing Periods

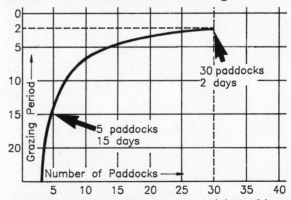

The curve shows how the grazing period derived from an average 60-day recovery period shortens as the number of paddocks rises.

may benefit the land and allow feed supplement cuts that justify 100 paddocks or more. Aesthetic and legal considerations, type of terrain, cost of labor, herders, and your three-part goal, of course, might override these factors.

The Effect of Paddock Numbers on Density

"Stock density" is the number of animals per unit of land (animals per acre in this book) at a given moment. Since it involves both animals and acres, it influences the ADA consumed from paddocks in a cell.

$$\text{Stock density} = \frac{\text{animals}}{\text{acres}}$$

The "acres" in this equation does *not* mean the total acres in the cell, but the acres in a given paddock. Thus you could have high cattle numbers and low stock density or vice versa.

Paddock Numbers and Stock Density

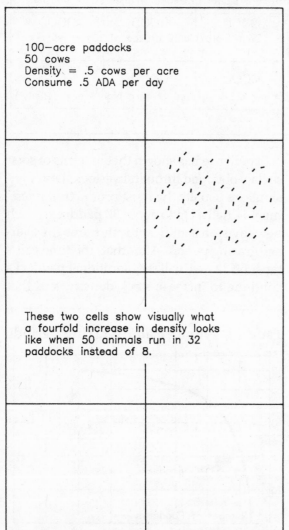

100—acre paddocks
50 cows
Density = .5 cows per acre
Consume .5 ADA per day

These two cells show visually what a fourfold increase in density looks like when 50 animals run in 32 paddocks instead of 8.

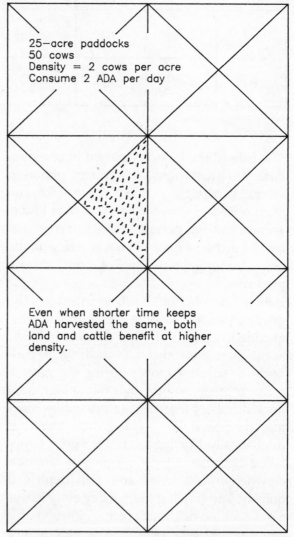

25—acre paddocks
50 cows
Density = 2 cows per acre
Consume 2 ADA per day

Even when shorter time keeps ADA harvested the same, both land and cattle benefit at higher density.

These two cells show visually what a fourfold increase in density looks like when 50 animals run in 32 paddocks instead of 8.

Even when control of time keeps ADA harvested the same, both land and cattle benefit at higher density.

Because stock density includes the same animal and acre factors we use to compute ADA, it also represents how many ADA a paddock will give up every day the herd is in it: Every acre in a paddock stocked at 10 cows per acre density will give up 10 ADA every day the herd spends in the paddock.

At high density, therefore, the impact on an acre in 1 day becomes very great—and extreme damage can result when herds spend too long in a paddock or return too soon. In the later case they may *severely* overgraze nearly every plant.

The advantages of high stock density are:

- Animals tend to graze a greater proportion of available plants and graze them more evenly, leaving fewer ungrazed or severely grazed.

- Distribution of grazing, dung, and urine becomes more even.

- Animals move more frequently onto fresh ground, stimulating them and providing a more constant level of nutrition.

- Tighter plant communities tend to develop, providing more leaf and less fiber.

- Animal performance improves.

(Note: If forage is growing and very rich, your animals' performance may actually drop due to lack of roughage. In this case you may have to feed them supplements of minerals or old hay.)

In your planning you'll probably consider increasing stock density in some areas before others and proceed by dividing existing paddocks or grazing areas. (See the textbook for details.) When you do this you'll notice the subtle relationship between stock density, paddock numbers, grazing periods, and recovery periods. The planning procedure at the end of this part (see "Creating Your Plan") explains the mathematical rules. The following example demonstrates the principle.

EXAMPLE

Effects of Splitting Paddocks

This cell has six equal paddocks. A grazing period of 12 days in each will allow each paddock 60 days to recover. The herd will cover the whole cell in 72 days.

Cutting paddock 1 in half to create paddock 7 will obviously double the stock density in these two paddocks. If, to keep the ADA constant, you then graze the small paddocks only 6 days, you'll notice that they now get 66 days to recover instead of only 60.

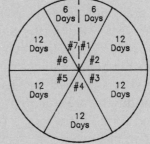

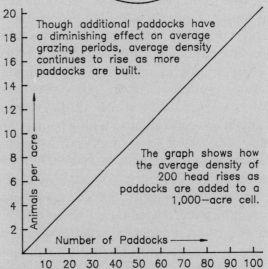

Though additional paddocks have a diminishing effect on average grazing periods, average density continues to rise as more paddocks are built.

The graph shows how the average density of 200 head rises as paddocks are added to a 1,000-acre cell.

In practice, for the sake of convenience, we compute the grazing periods from a desired *average* recovery period rather than from a minimum like 60 days in this example. The numbers don't look quite so neat, but the principle remains: When you subdivide paddocks, you reduce grazing periods and increase stock density and recovery periods in those paddocks.

Forage Reserves

Traditional practice calls for keeping some land in reserve to carry animals in time of drought, fire, or other catastrophe. It's reassuring to look over at tall grass waving in the lower 40 and know it's there if you need it.

On the other hand, owing to natural processes explained in the textbook, forage simply left standing for long periods of time loses nutritional value and is easily lost to fire. Overrested land declines in productivity. And if you don't use the reserves, animals will select new grass and leave the old to weather yet another year. Also, withdrawing some land from production puts heavier pressure on the remaining land and on animals. The long-range effects are bad, especially when drought conditions are building.

Compared to the extra forage production possible when planned grazing encompasses all the land, withholding land for drought reserve typically results in a net forage loss and lower animal performance. This practice persists mostly as a defense against human nature. You know that in theory you need only one bank account, but putting something in a second account with a penalty for early withdrawal helps suppress the temptation to dip into savings. A pasture withdrawn for drought reserve serves the same function.

Since holistic planning offers a way to compute how many animal-days a whole cell can supply, you can budget a certain amount for reserve. Then you can handle the accounting separately on paper, just as the bank does, yet manage total capital, *on the land*, in the best interests of plants and animals.

The example to the right demonstrates the mathematics of reserving area verses reserving time, and shows how plans that may be equal in terms of forage consumed result in quite different treatment of the land. Having extra paddocks available also increases management flexibility.

Holding Reserves in Time or Acres

Compare different reserve strategies for similar 1,000-acre, 10-paddock cells.

In terms of forage consumed, the plans seem equal.

Cell A grazes 800 acres and saves 200 acres for times when drought causes insufficient winter feed.

In a *good* year, the 800 grazed acres should yield:

$$\frac{200 \text{ cows} \times 365 \text{ days}}{800 \text{ acres}} = 91.25 \text{ ADA}$$

The reserved feed will be roughly 200 acres x 91.25 ADA = 18,250 ADs.

The yield could be lower as it is only a dormant season yield when finally used.

Land in Cell B will yield:

$$\frac{200 \text{ cows} \times 365 \text{ days}}{1,000 \text{ acres}} = 73 \text{ ADA}$$

Assuming grass grows equally in both cells, each acre of the time-reserve cell will have roughly:

91.25 – 73 = 18.25 AD of feed left

Total reserves are again:

1,000 acres x 18.25 ADA = 18,250 ADs

In a bad year, when reserves are necessary, all the numbers will be smaller, but the arithmetic will still show both plans equal.

Holding Reserves in Time or Acres, cont'd.

A. Acre-Reserve Cell

100 acres	100 acres	100 acres		
100 acres	100 acres	100 acres	100 acres	100 acres

B. Time-Reserve Cell

100 acres	100 acres	100 acres	100 acres	100 acres
100 acres	100 acres	100 acres	100 acres	100 acres

1,000 acres, 10 paddocks, 200 cows, 210-day growing season

In terms of production, the time-reserve cell will do much better:

1. Cattle in the acre-reserve cell take 25% more ADAs from the land they graze, meaning more grass starts recovering from lower on the growth curve (see page 44). Thus it regrows less in the same time than grass in Cell B.

2. Given an average 63-day recovery period in each cell, grazing periods are 2 days shorter in Cell B.

$$\text{GP time} = \frac{63}{10 - 1} = 7 \text{ days (Cell B)}$$

$$\text{GP time} = \frac{63}{8 - 1} = 9 \text{ days (Cell A)}$$

3. Cattle move to fresh grass 25% more often in Cell B, converting more of it into beef.

A simple technique allows you to estimate at any point how many animal-days of forage you have on hand if growth should stop right now. Applied another way, the same technique will tell you how many ADA land in the cell will have to supply in order to carry your animals for a given time. Then you can evaluate sample areas against that figure and look far down the road as droughts develop. Even more important, it allows you to plan for predictable nongrowing seasons, such as winter.

These calculations do not involve stock density at all, but rather stocking rate. Density, remember, refers to the concentration of animals in a paddock at a particular moment. *Stocking rate* is the number of acres you ask to support one animal for the time you expect a herd to remain in the cell as a whole. In other words:

$$\text{stocking rate} = \frac{\text{acres in the whole cell}}{\text{animals in the whole cell}}$$

Here's an example: If you have 100 cows in a 10,000-acre cell continuously, that is a stocking rate of "1:100—year round." A similar cell that carried 100 yearling steers only through the summer would have a stocking rate of "1:100—growing season only."

In planning, all stocking rates are handled similarly—except that when a cell supports livestock only part of a year, you'll have only the animal-days required by wildlife to worry about the rest of the time. The example on page 52 illustrates how to estimate the proper stocking rate for any piece of land.

It takes a lot of experience to decide what size square of land will support an animal for a day—and you must check several places on your land because there may be big differences between sites. Records for past years will tell you what your land really does yield. Monitoring and fixed-point photos will also give you a feel for changes from year to year.

EXAMPLE

Figuring Approximate Stocking Rate

Imagine a sheep ranch with a stocking rate of about 1:20. That means 20 acres for every sheep. To keep the numbers simple, assume you're keeping no lambs through a winter which lasts from mid-October to mid-April (180 days).

Thus every 20 acres has to supply 180 sheep-days of winter feed:

180 sheep-days ÷ 20 acres = 9 sheep-days per acre (ShDA)

So 1/9 acre must be able to feed one sheep for one day.

Now look at your land after first frost and see if one sheep could eat for one day on a selection of 1/9th-acre samples.

1 acre = 4,840 square yards
1/9th acre = 4,840 ÷ 9 = 538 square yards

Pushing the square-root button on a pocket calculator will quickly tell you that one side of a square covering 538 square yards is 23 yards. You can then step off squares and decide if each square has enough feed to carry a sheep for a day—remember that wildlife also take some.

You can also work the same problem the other way to find out what stocking rate your land *can* carry through the winter. After a very dry summer, suppose you find that a square of 23 yards to a side will not feed one sheep for one day. Suppose you find that the square has to be at least 35 yards on each side. From this you can determine how much stock you must sell to get through the winter (or how many sheep-days of feed you must buy):

35 yards x 35 yards = 1,225 square yards

One acre (4,840 square yards) supplies 4 sheep-days (4,840 ÷ 1,225 = 3.95).

A 40,000-acre ranch supplies 160,000 sheep-days (4 ShDA x 40,000 acres = 160,000 sheep-days).

160,000 ShD feed only 889 sheep for 180 days (160,000 ShD ÷ 180 days = 889 sheep).

If you normally run 2,000 animals, you can expect them to run out of food on the range in about 80 days (160,000 ShD ÷ 2,000 sheep = 80 days).

To figure an animal-day plot

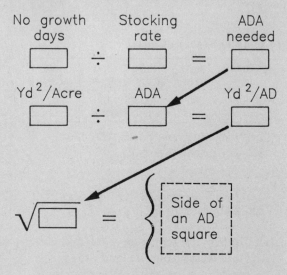

To figure ADs available

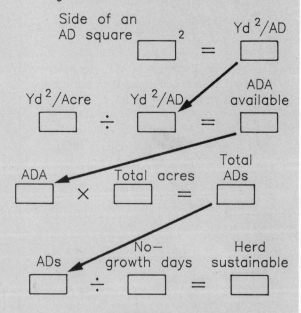

Roughly, however, if you can imagine yourself successfully filling a sack the size of an animal's stomach or making half a bale of hay while picking with only one hand, you are in the ball park.

In practice, when you go out on the land and lay out sample squares to see if your forage will last the winter, if all is well, there should be no doubt. If you find yourself fudging a bit and dragging your feet on testing areas that you suspect fall below the average, you're probably in danger.

Then you have the challenge of working the second problem—deciding just what size square *will* carry an animal for a day. This gets pretty tricky. Even though the procedure outlined here provides the best information you are likely to get, it's still very crude. A few yards can make a huge difference in your calculation of total reserves, so be prepared for some real soul-searching and don't be afraid to get multiple opinions.

A Final Word About Stocking Rate

The technique of measuring forage reserves by checking the area needed to feed one animal for one day can also help you during the growing season to determine the limits of stocking rate. For the next paddock you intend to graze, compute the size of the square necessary to feed one animal for one day during the shortest grazing period allowable for that paddock. Because you would use that square only during ideal growing conditions, you are certainly overstocked if that size square doesn't pass the test.

Example: Suppose you plan 800 cows for a 600-acre paddock on 3-day to 9-day grazings.

800 cows x 3 days ÷ 600 acres = 4 ADA

4,840 square yards per acre ÷ 4 ADA = 1,210 square yards per animal-day

Square root of 1,210 square yards = 35 yards (one side of an animal-day square)

If this passes, compute the smaller square that would support one animal for one day, given the longest grazing period in that paddock (9 days in this case); see page 54. If that doesn't pass, you're probably gambling on rapid growth. Monitor carefully.

Estimating the area required to feed one cow for one day.

If animals seriously deplete forage or start to eat litter before they're scheduled to move, suspect overstocking but check other factors:

• You might have overestimated a paddock's capacity. Taking more ADA from a better paddock might relieve the pressure.

• Low stock density can produce forage too rank to eat. Mowing, burning, or herd effect can solve that problem, but plan for higher density.

Longer Grazing Periods Mean Smaller Sample Areas

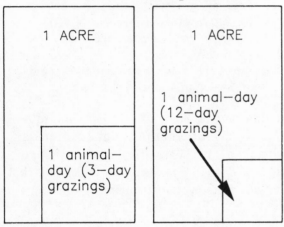

The longer the grazing period, the smaller the square that must provide each animal-day of forage.

In year-round grazing situations, overstocking usually takes its toll only when animals exhaust forage during the dormant season. Thus if overgrazing in the growing season was minimized, overstocking hurts animals, wild and domestic, more than it affects plants or the land. Nevertheless the effect is bad. Consider:

• As reserves dwindle, grazing animals will eat litter, reducing the soil cover important in all four ecosystem processes.

• Scarcity of plants tall enough to penetrate snow harms species such as pronghorn

that do not forage through snow. (Plants that penetrate snow cause it to melt faster and may actually open up patches of grass. They also hold blowing snow, returning more moisture to the soil.)

• Certain animals will shift exclusively to browse, creating "browse lines" below which younger, shorter animals can find nothing.

These problems appear most frequently in yearling operations that do not plan to keep livestock through the dormant season and do not budget a winter reserve for wildlife. Even so, year-round grazing plans must recognize the danger, too.

If you do intend to manage for any significant wildlife population at all (remember, diversity is almost always an asset), careful planning of forage will affect game populations more than any single factor, including hunting. Forage governs how many wild animals survive the dormant season and where they spend it. An elk herd breaking fences to get to a haystack can cost plenty.

Some stockmen at the Los Ojos Ranch in eastern New Mexico invented a fairly precise way to check their estimates. They subjected one paddock to the grazing pressure they figured all their land would have to stand, on average, during the winter, but they fenced *out* some sample squares representing what they estimated would have to feed one animal for one day at that level of use.

At the end of the test period they could look at "before" and "after" squares and know how close they came. If the ground that was grazed still contained acceptable litter and forage, they could expect that similar use through the winter would benefit the land, and cattle performance would not suffer. Then they could use the fenced-off patch as a proper example of a square that could indeed feed one animal for one day. The Los Ojos people rejected the idea of putting one

A cow-day enclosure.

heifer in a fenced-off square on the grounds that she wouldn't graze normally under those conditions, nor would the ground get the treatment it would from a complete herd.

The Critical Dormant Season

An underlying axiom of holistic management holds that naturally functioning wild herds do not destroy land. Animals may die off occasionally because of drought, late springs, and other climatic surprises but almost by definition these setbacks occur when plants are not growing and the danger of overgrazing is minimal. Thus wild population levels reflect the amount of forage available in the dormant season.

For domestic stock also, it is the amount and quality of standing forage available at the onset of the dormant season that largely determines your stocking rate. No other factor tells as much about the carrying capacity of your land.

If, as in a yearling operation, your range doesn't have to carry a herd through the dormant period, regulate your numbers to leave enough animal-days of forage for wildlife and some reserve against a late return of growing conditions. Otherwise your numbers will reflect what you can carry through the winter plus wildlife and emergency reserves.

Only in the most extreme cases of overstocking combined with bad timing do animals run out of feed during a growing season. Far more typically, a dry summer does not leave enough standing forage to last through the winter. In America trouble hits about February. Sometimes drought problems are just as predictable. In northeastern Arizona, for example, ranchers can expect some moisture in March or April, but if mid-May arrives dry, they had better figure out what their stock will eat and drink until August. The odds get very long in June and July.

Whenever you think you may run out of feed during a time of minimal or no growth, keep these points in mind:

• Before cutting stock, use the Los Ojos technique described here and see how many animal-days of forage you have. If you really *must* destock, the sooner you know it and act, the less you have to cut.

• Consider changing plans to enhance any growth that plants do achieve and amplify the rebound when good conditions return. This generally means combining stock into fewer, larger herds so paddocks get more intense but less frequent use.

• Look at ways to use available forage in the most efficient way possible.

The Economics of Destocking

At HRM training sessions, hundreds of professional land managers have faced this question: "If you have to destock, why should you do it right now, not later?"

Most people just resort to prayer until the hand of God forces the issue, but in a classroom they think up a reason fast. Ninety percent say, "So you can sell before the price falls." The rest usually split between "Perfor-

mance may drop if you wait" and "You'll beat hell out of the ground."

Very few ever see the main point: If you have to destock, the sooner you do it, the less you have to cut. The accompanying graphs illustrate why this is true. They describe what happens to 500 cows in a 160-day winter season when forage reserves fall 25% short. Normally the herd would consume 80,000 AD during that time. What plans make most sense when you estimate you'll have only 60,000 AD standing at the onset of cold weather?

This example is extreme because in reality mature animals don't often starve to death. They just abort, fail to conceive, shrink down to scratch, get hauled to the sale barn while they can still stand, or dine on the benefit of emergency loans. Nevertheless, forewarned still proves forearmed.

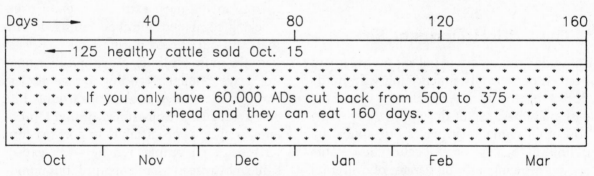

500 animals x 160 days = 80,000 AD   but 60,000 AD ÷ 160 days = 375 animals

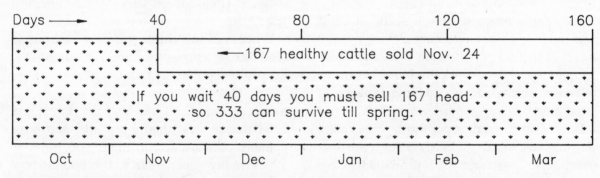

60,000 AD − (500 animals x 40 days) = 40,000 AD ÷ 120 days left = 333 animals

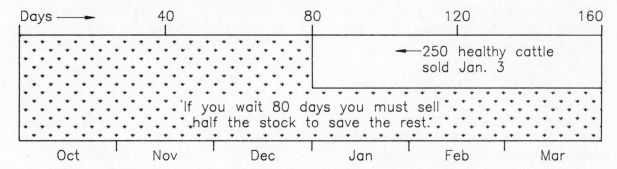

60,000 AD − (500 animals x 80 days) = 20,000 AD ÷ 80 days left = 250 animals

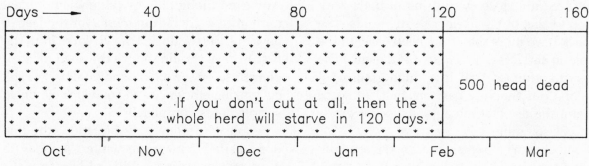

60,000 AD ÷ 500 animals = 120 days before everything is gone

Nutrition During Dormant and Slow-Growth Periods

Earlier we looked at the advantages of greater stock density during the growing season, which you can get by combining herds and cells into one larger cell with more paddocks. A dormant season poses a slightly different problem.

Let's say you have enough standing forage after the first freeze of autumn to last you till spring, including plants that cure well enough to supply nearly all the protein and vitamins your livestock and wildlife require. Nevertheless, though all animals can find enough bulk throughout the season, its nutritional value drops steadily under continuous grazing. Both wild and domestic animals will select the best first, getting more than they need during the first half of the winter and

less during the second. You can buy supplements to make up the difference for the livestock. Wild grazing animals will go short.

Other nasty things can happen too—such as increased poisonous plant danger in the spring, as many of the most dangerous species green up early. For the same reason, cool-season grasses and forbs suffer intense early grazing and frequently overgrazing, as in milder climates many of them actually go dormant only a very short time.

On ranges with higher successional grass communities, the better-curing grasses often give a reddish hue to an autumn landscape. By late winter, however, that color not uncommonly pales to the straw yellow of less nutritious plants that remain. In late February in the American Southwest, semiferal horses sometimes wander the roadsides through thick stands of pale old grass looking for cast-off Pampers to eat. The problems tend to be worse on high-rainfall leached soils as even growing vegetation offers less in protein and minerals and more in fiber, and it cures poorly and decays quickly.

Graph A above assumes supplementation will be necessary after exactly half the forage has been selected (and the need will increase). And the stock, being scattered, fouls a wide area, causing inefficient grazing and never allowing the stimulation of a move onto fresh ground.

Real situations are never neat, of course, and require close monitoring of both stock and forage, but the principle still holds. The stock have more than they need early in the season and less than required at the end. The mining industry calls this "high grading"— taking out the very best ore first. At some point the ore that remains is too low-grade to pay for mining it.

Miners can't make any more ore, just as you can't grow much grass in winter. But by mix-ing high and low grades, they can economically mine much more. The more paddocks you have, the better you can achieve the same thing.

Graph B shows the effect of using 10 paddocks one after another. The cattle spend 20 days in each paddock over a 200-day dormant period.

This scheme has several advantages. Stock do occasionally move to fresh ground. The ungrazed paddocks will provide something for wildlife right to the end of the season. And even though forage value will not meet minimum daily requirements for the last 10 days in each paddock, the carryover from the first 10 days should reduce the need for supplements. On the other hand, fouling will still cut grazing efficiency and you won't eliminate supplements completely.

People who follow a rigid rotation schedule don't usually change in winter. A 3-day to 4-day grazing period would look like Graph C.

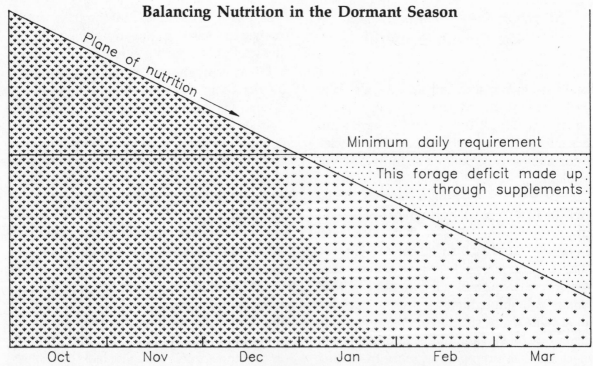

Balancing Nutrition in the Dormant Season

Plane of nutrition

Minimum daily requirement

This forage deficit made up through supplements

Oct   Nov   Dec   Jan   Feb   Mar

Graph A: Continuous grazing creates a large nutritional deficit at the end of the season.

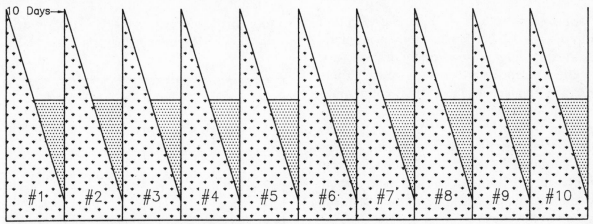

Graph B: Even a few paddocks will somewhat balance nutritional needs across time.

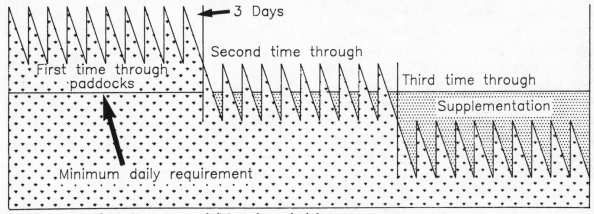

Graph C: A rapid rotation creates a deficit at the end of the season.

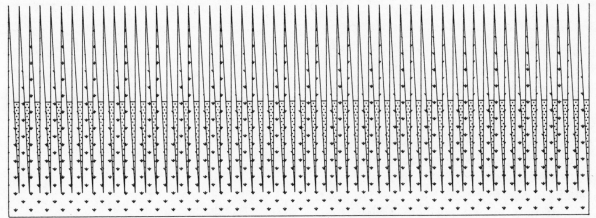

Graph D: High paddock numbers keep nutritional plane high throughout.

This scheme would in fact keep stock moving onto fresh ground and greatly reduce the fouling problem. You might get by without supplements through the second rotation. You would not meet the requirements of wildlife, however, and the supplement bill during the last rotation would be high.

Suppose, however, you have 100 paddocks and the herd stays only 2 days in each. Then the pattern looks like Graph D.

Your stock move to fresh ground every other day. Forage use becomes extremely efficient. And supplements are virtually unnecessary.

These examples are of course purely theoretical. Other considerations may count much more in your planning. Some paddocks may remain snow free. Some may provide cover or a convenient place for monitoring calving or lambing. Nevertheless, amalgamating herds in the dormant season means more paddocks, more density, better nutrition, and greater efficiency.

The policy of reserving time (in animal-days) versus reserving area (in ungrazed paddocks) merits another look. If drought or a late spring forces you to make an extra pass through all your paddocks, your stock will face a depleted nutritional selection in contrast to the reserved-area policy. Nevertheless, the policy of reserving time almost always proves better. Using all the paddocks will produce more reserves and better livestock condition at the point where reserves become necessary—and rapid moves onto fresh ground will stimulate performance.

Amalgamation of Herds in Time of Drought

Planning in time of drought differs slightly from planning at the start of winter. Not only is there no event such as first frost to define the beginning, but the end is not predictable. Growth slows down drastically but only stops when conditions become truly extreme.

At the first pang of worry about the prospects ahead, estimate your animal-days of standing forage to get an idea of how long you can last in a "worst case" scenario. Then weigh your options—buying hay now, destocking, leasing pasture, and so forth. But if you normally run several herds in separate cells or paddocks, you have another possibility: combining herds.

The greatest obstacle to reaping the benefits of combining herds is worry about handling

How Joining Herds Affects Grazing and Recovery Periods

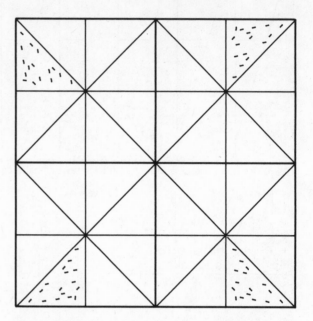

Four herds of 125 cows in four 1,000-acre, 8-paddock cells. Density is 1 cow/acre. A 14-day GP yields a 98-day recovery period.

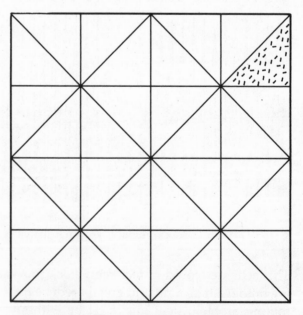

Cells combined into one 4,000-acre, 32-paddock cell. Density is now 4 cows/acre. A 3-day GP gives a 93-day recovery period.

> *Combining herds and treating several cells as one cell gives land longer recovery periods and additional benefits from greater stock density. It may also help you maintain a constant level of nutrition, thus reducing the need for supplements.*

so many animals. Many experienced people vow that all kinds of problems arise when numbers top 140 head.

Specialized breeding programs and other considerations may make separate herds

The Barlite Case

On the Barlite ranch near Marfa, Texas, managers Charles and Katie Guest faced reduction of the 1,200-head herd they had divided among seven cells, containing a total of 101 paddocks. They estimated their reserves and figured they could survive if each square yard of ground merely grew an additional half ounce of feed.

To maximize animal impact and minimize the risk of overgrazing, they put all 1,200 head together and moved them daily. This gave each paddock a maximum dose of dung, urine, and trampling, followed by 100 days of recovery.

"I didn't think anyone could move that many cattle every day," Charles Guest remarked later. "But they were so used to the fences already, they pretty well moved themselves."

Finally in October and November an inch and a half of drizzle blessed the Barlite. The Guests figure it grew them 21 million pounds of feed. The sudden lushness of their ground stopped abruptly at their boundary fence—beyond which their neighbors' cattle had grazed continuously at half the Barlite's stocking rate. While the neighbors continued to destock, the Guests bought 206 cow-calf pairs at distress prices and cut their supplemental feed bill by $26,000.

necessary, but simple numbers usually don't. The doubters generally do not believe that animals can learn behavior that makes herd size almost irrelevant to the question of handling. In fact nobody has ever proved any upper limit, though no doubt every situation has one. If you do have the option of combining herds in a flexible way, consider now how you could use the nutrition available in standing forage more efficiently.

Herd Effect

Herd effect—the hoof action of excited animals on plants and soil—is perhaps your most powerful tool in managing succession in brittle environments. Whereas stock density, another key aspect of animal impact, is a mathematical relationship between the number of animals and the size of the grazing area, herd effect is only a matter of behavior. Theoretically a herd of any size can produce it on any piece of land. But:

> *The bigger the herd, the better the herd effect.*

This is not a linear relationship. A herd of 1,000 can generate much more than 10 times the amount of herd effect produced by 100 head. Very small herds will not create much herd effect at all.

In biological planning, the idea is to anticipate the areas where you will apply herd effect for any number of purposes, including the following:

- To suppress brush directly by breaking it down

- To return stale, ungrazed plant material to the soil as litter

- To promote succession toward grassland or tighter spacing between plants

- To soften the banks of gullies and start succession in eroding areas or cropland being returned to pasture

- To reduce infestations of noxious weeds by direct impact and by creating soil conditions that favor fibrous-rooted grasses and sedges over tap-rooted species

- To clear firebreaks or roadsides

By withholding herd effect, you can promote brush in areas where you might want it for wildlife habitat, winter cover, and the like.

In the wild, predators account for a large degree of herd effect. In fact, game animals as well as domestic stock tend to become placid when free of that danger. Driving livestock with cracking whips or dogs obviously causes herd effect but at an unacceptable price in lost performance and handling qualities. Positive inducements, however, do not have these side effects.

For example:

- Supplements such as hay or cake fed on the ground will quickly excite any herd trained to expect a handout.

- Salt will gather a herd that has been denied it for some time. Granulated livestock salt, simply fed on the ground, works best.

- Diluted molasses sprayed on weeds or firebreak areas will stimulate both grazing and herd effect on specific locations.

- Static inducements such as salt blocks and liquid mineral licks do not produce herd effect. Animals visit them singly and tend to loiter. Putting mineral supplements on a trailer that can be moved from place to place works better but falls short of the ideal.

Training plays a large role in all these techniques. Animals that have never tasted molasses, for instance, will not recognize the smell and may ignore it at first. Livestock will quickly learn to come to a whistle, though, if it consistently means a treat. Such training not only helps in stimulating herd effect but also simplifies the business of moving stock to new paddocks or grazing areas. Holding back a few trained animals to mix in with untrained stock vastly speeds this training.

Multiple Herds

Although the land in a cell benefits most when livestock run in a single herd, many situations call for running two or more herds separately. You can do this in three ways:

- Assign several paddocks to each herd and plan each division as a subcell.

- Move separate herds among all paddocks while keeping recovery times adequate.

- Have one herd enter a paddock as another leaves ("follow-through grazing").

The planning procedure in the next section ("Creating Your Plan") tells how to compute grazing periods—but they may prove unacceptably long in cells with few or very unequal paddocks. Although follow-through grazing is particularly tricky to plan, it does fill certain needs best:

- When herds require different levels of nutrition (say, first-calf heifers and mature cows)

- When different types of livestock impact forage differently (goats following cows may use browse better)

- When topography or labor considerations favor keeping herds close together

The procedure on page 74 tells how to compute grazing periods for each herd on follow-through grazing. The diagram shown on the next page presents two cases. Equal paddocks cause no problem, though plants are exposed to animals for twice the grazing period of one herd. Putting two herds in Cell II could involve some heavy mathematics.

In practice the following guidelines plus careful and continual monitoring will usually serve:

- If one herd is small, you might key all moves to the larger herd. Holding a small number longer than desirable in smaller paddocks will not hurt much. (But still calculate the ADA.)

- If the moves progress from larger to smaller paddocks, the following herd can skip an occasional paddock and catch up.

- If moves progress from smaller to larger paddocks, the lead herd can skip ahead, while the following herd actually makes the first grazing of the intervening smaller paddocks.

Note that when one herd skips a paddock, you shorten recovery times. So try to plan similar paddocks in sequence.

Matching Animal Cycles to Land Cycles

Stockmen may argue forever about the best time of year for calving and lambing. Many variables govern the decision—markets, parasite hatches, seed and awn problems, weather, and more. That said, however, tradition often outweighs most other factors. If your production goal is profit and you aim to lessen the burden of supplemental feed, low conception rates, and care of young stock, checking your herd's varying nutritional demands against the natural cycles of forage might prove interesting.

Traditional practice usually dictates that calves come in early spring to be weaned and sold in the fall. If the market looks bad, some may be held over till the next spring. But if other factors (say the absence of screwworms) permit it, why not consider summer calves? Generally the economics of holding them over and selling them at 18 months look

1	2	3	4	5	6	7
180 acres	180 acres	180 acres	180 acres	180 acres	180 acres	180 acres
12 Days	12 Days	12 Days	12 Days	12 Days	12 Days	12 Days

Cell I: Follow-through grazing with 12 days in equal paddocks gives 60 days recovery.

1	2	3	4
140 acres	140 acres	140 acres	280 acres
	15 Days	15 Days	30 Days
15 Days	7	6	5
	140 acres	140 acres	280 acres
	15 Days	15 Days	30 Days

Cell II: The follow-through plan below gives 60–105 days recovery in unequal paddocks.

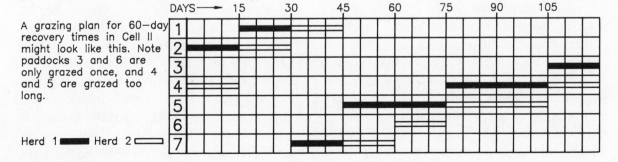

A grazing plan for 60-day recovery times in Cell II might look like this. Note paddocks 3 and 6 are only grazed once, and 4 and 5 are grazed too long.

Herd 1 ▬▬ Herd 2 ▭

EXAMPLE

Deciding the Number of Herds

You have 600 steers on 2,000 acres divided into 36 almost equal paddocks. You'd like to push 100 of them ahead and market them early. Consider 3 strategies:

1. Keep all 600 together in 1 herd, and select the best at sale time.

2. Put 100 in the 10 best paddocks and 500 in the other 26.

3. Let the 100 make the first selection in each paddock, and let the 500 remaining follow right behind.

If you desire a 60-day average recovery period, the following factors will bear on your decision: length of grazing period, ADA harvested, and stock density. Consider:

-   Land benefits from larger herds, high density, and short grazing periods.

-   Livestock benefit from rapid moves to fresh ground and lower ADA take.

To choose, think first about impact on land, then on livestock. Thus:

-   Plan 2 does much less for the land because of longer grazing periods, unequal ADA harvest, and low density in 10 paddocks.

-   Because of fairly high paddock numbers, Plans 1 and 3 have a similar impact on land.

-   Plan 3 will better handle your livestock performance goals.

Plan	Herds	Paddocks	Average Grazing Period (days)	ADA Each Grazing	Stock Density
1	1/600 head	36	1.71	18.46	10.8/acre
2	1/100 head	10	6.66	11.98	1.8/acre
	1/500 head	26	2.4	21.6	9/acre
3	2/100/500 head	36	1.76/1.76	3.17/15.84	1.8-9/acre

much better than for calves dropped in early spring.

Once again, the standard worksheet provides a handy form that allows you to compare several strategies or classes of stock and different kinds of country. Once you start considering possibilities, the decisions can become extraordinarily complex.

Roughly speaking, cows need a rising plane of nutrition for conception and above-average feed for 6 months of lactation. They need less when dry and pregnant. Sheep and goats follow a similar pattern except that they're usually not lactating and pregnant at the

same time. The accompanying graphs show the general nutritional needs of cattle and sheep.

Rethinking the cycle of forage may even turn up the fact that keeping stock through the winter doesn't pay at all. Or you might want a combination of operations so that market times support a constant flow of cash.

The Deseret Ranch, a 400,000-acre operation in Utah, runs yearlings, a cow/calf operation, a herd of bison, and a lucrative elk-hunting enterprise. Their country, in several locations, runs from nearly snow-free to alpine. The bison forage well through snow,

Nutrition and Forage Cycles

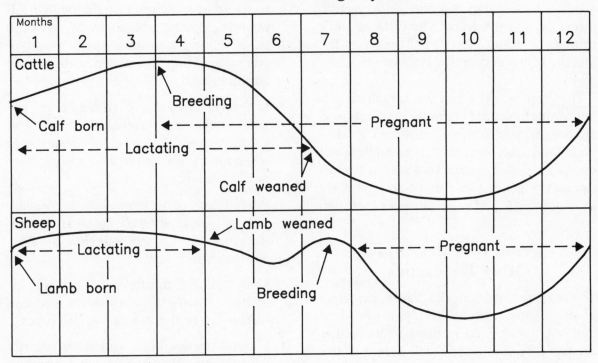

Forage Availability

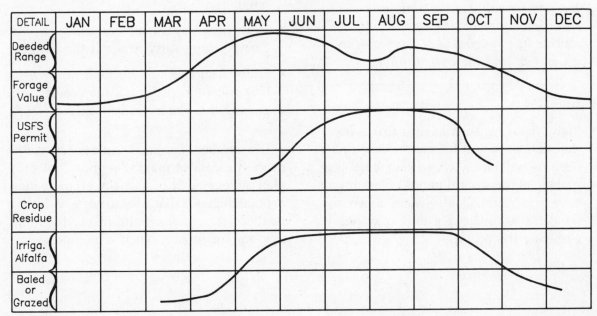

Questions: Summer calves and early lambs or ewes and yearlings on permit land? If south slopes and low-elevation parks were not grazed late in summer, would elk winter there instead of on alfalfa? Could we sell a better hunt? Could we substitute grazing for last alfalfa cut? Wean lambs onto residue? Finish them on alfalfa? Registered fall calves? Spring lambs and stockers?

but the ranch also cuts hay and has some significant plantings of basin wild rye that grows tall and stiff enough for cattle to graze through the snow. Even a small operation can become nearly as complex, once you consider all the options.

The point is that when you actually begin to do your biological planning for plants and animals, no tradition need be sacred. And as your land improves, new possibilities will open—for instance, more cool-season grasses extend the grazing season, poisonous plant problems fade, and water supplies increase.

Pests, Parasites, and Other Headaches

The sequence of paddocks grazed can affect a host of considerations besides the productivity of plants. Most of these relationships are too case-specific to detail here, but the principle is simple:

You can probably manipulate any situation characterized by a strict routine by changing your own routine, if you can find the critical point.

Examples:

• Many parasites such as liver flukes leave their livestock host for part of their cycle. Records will show where your herd was grazing at this critical period. Planning to have your animals elsewhere when the parasites again need a host may nearly eliminate the problem.

• Some pests either breed in manure or seem to spend a lot of time on moderately fresh manure. Moving animals frequently onto fresh ground will keep them ahead of emerging young flies and may leave some adults behind.

• Livestock losses rise when predators have good cover, insufficient wild prey, and young to feed. Plan to calve on safe sites when natural prey is plentiful and predator needs are low.

• When floods occur in predictable seasons, plan to graze affected areas afterward, since ungrazed plants will slow the flow and catch sediment.

• Many poison plants threaten stock only for limited periods or when other forage is scarce. Graze these areas at safe times.

• Ground-nesting birds and other animals may require certain habitat only for nesting or breeding. Keeping stock out of these areas at critical times will increase their numbers.

• Grass fires may threaten certain areas, but grazing them early may cut the danger.

• Shifts in succession (desirable or undesirable) in certain paddocks may come from grazing them at a particular time.

The number of options that must be sorted in the process of making a good biological plan can obviously overwhelm even the most able intelligence that tries to cope without a method for organizing and displaying information. Fortunately, such a method exists.

CREATING YOUR PLAN

Biological planning is done according to an *aide memoire* (French for memory aid) because, as the textbook explains, the task involves more variables than the average mind can handle simultaneously. As the term applies here, an *aide memoire* differs from a simple checklist because it gives a sequence for making decisions that takes into account the effect of one decision on another. Order usually doesn't matter on a checklist. The *aide's* main benefit is peace of mind, as it allows you to limit your thinking to decisions your brain can handle. You can concentrate on one step at a time without worrying that something else should have come first. When you complete each step, you wipe it from your mind and focus on the next. Nothing gets left out.

The years of experience embodied in the *aide memoire* assure that you will incorporate all concerns you may have, even if they don't turn up in the order you anticipate. Do not skip ahead. Along the way you will decide where to apply the tools of rest, grazing, and animal impact according to the guidelines of population management, time, stock density, and herd effect. Other steps will enable you to plot the application of other tools and uses of the land. As you constantly monitor your progress, you'll be able to modify and replan according to changing circumstances.

Where sophisticated day-to-day management may not be available—often in communal/tribal situations—we suggest you use a set of guidelines, in lieu of the *aide memoire*, that have been developed for these instances. These will assist in planning the concentration and movement of herds as befits various situations, so that land improves and stocking rates can rise. Individual animal performance, however, may not reach its full potential, and wildlife needs and other land uses will benefit only indirectly from the general improvement in productivity.

The *Aide Memoire* for Biological Planning will generate a plan that takes into account any number of variables and displays them in a graphic way and in great detail. Thus you can advance several priorities at once, see your situation at a glance, and change the plan easily as circumstances decree. As the simplified guidelines are adapted from the *Aide Memoire*, we present the latter first.

Warnings

As in the case of the financial plan, the instinctive human fear of planning in general is your greatest enemy. More than you wish to admit, however, this fear reflects the old "what you don't know won't hurt you" syndrome. Planning will show up areas where you can't attain the ideal—and that hurts.

Remember specifically:

- Follow the *Aide Memoire* carefully. Do not skip steps or take them out of order.

- Plan on paper. The number of factors will overwhelm even the best unaided mind, and others cannot contribute much to a plan they can't see.

- Plan pessimistically on any point on which you have any doubt at all.

- Don't take shortcuts. Use the planning chart and complete all the rows and columns your plan calls for.

- Plan creatively every time. Easy planning year after year will tempt you to abandon the process and fall into a routine that will sooner or later lead you far enough from your goals to cost you plenty.

- Don't abandon common sense. What you put on paper is a guide—your most educated guess about the future. It is not a law you cannot bend or change when reality dictates.

The *Aide Memoire* for Biological Planning

Step One:
Opening Decisions

Choose the Season

Holistic management requires you to plan, monitor progress continuously, control deviation as soon as possible, and replan whenever necessary. Even though this plan—monitor—control—replan sequence proceeds without gaps and covers emergency situations, livestock operations usually call for major planning twice a year.

Make your first plan at least a month before the onset of the growing season. It should project well ahead but remain open-ended as you can't predict precisely when growth will end or exactly how much forage will grow before that date.

Make the second plan toward the end of the growing season when you know the forage reserves available during the non-growth period. Because the amount of forage will not change, you can project the plan forward to a theoretical endpoint. As landscape

and production goals will suffer from total consumption of forage and litter, this "closed plan" should cover the normal dormant period plus a prudent allotment for wildlife and drought reserve and still leave a comfortable margin.

Picture the Grand Design

Gather for a brainstorming session all the people responsible for putting the plan into effect. Use the model of brainstorming described on page 16 and cover all possible factors and influences touching livestock, wildlife, crops, haying, and the landscape goal. Don't judge any idea or go into detail. Just write down all the ideas and keep the list handy.

Next focus on the time of year you're planning for and try to envision the whole ranch/farm. Then answer the following questions:

- What areas do you intend to plan as cells? (A cell, remember, is any group of pad-

Open-ended and Closed Plans

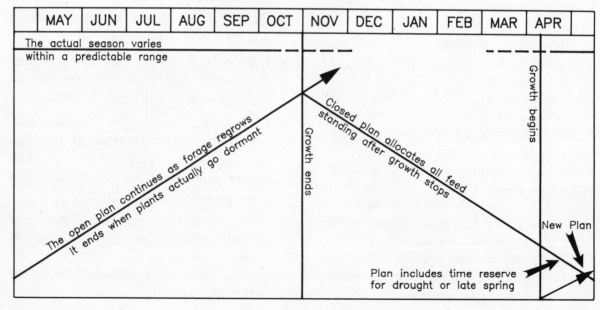

River Bench Ranch

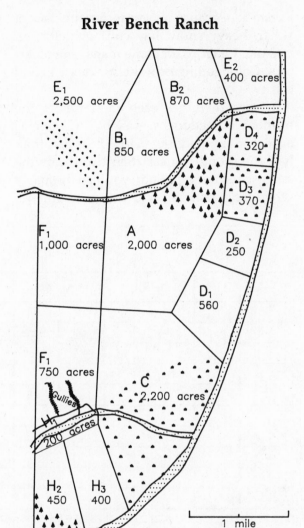

E₁
2,500 acres

E₂
400 acres

B₂
870 acres

B₁
850 acres

D₄
320

D₃
370

F₁
1,000 acres

A
2,000 acres

D₂
250

D₁
560

F₁
750 acres

C
2,200 acres

Gullies

H₁
200 acres

H₂
450

H₃
400

1 mile

Rank old grass ⬚

Loco weed ⬚   Storm cover ⬚

A cell (often a whole ranch is planned as a cell) is usually complex. This example is used in completing the planning chart illustrated on the pages that follow.

docks or unfenced grazing areas that you plan as a unit to control time and thus minimize overgrazing. It can contain more than one herd and change from year to year.)

• Do you want to run herds just as before or change?

• Did any areas receive more rain than others or change in any significant way?

• What stocking rate do you intend to carry? (This could change in the course of the planning, but at this stage you'll probably have an idea of what you want and what you think is reasonable.)

• What crops will be planted, and where? (If these are planned on worksheets, you can see clearly where livestock run or are needed on crop fields.)

Step Two: Setting Up the Planning Chart

Cells are planned on the Biological Plan and Control Chart illustrated on the next page and in Appendix B (blank copy).

Record Basic Statistics

Use one planning chart for each cell that you plan as a unit. If you have more than 20 paddocks or grazing areas, cut apart other charts and paste on enough sections to give every paddock a line. Then:

• Record the paddock numbers and sizes in column 3.

• Enter the total cell size and stocking rate in row 36.

Generally it's helpful to have the main growing months in the middle of the chart. If you live in the Southern Hemisphere, you may want to reprint the charts with the calendar starting at July rather than January, which would move the growing season to the middle of the chart.

Record Management Concerns

Using color-coded felt pens, box in all the management events that affect the cell as a whole. Draw vertical lines through all the paddocks on the starting and ending dates and connect them across the top. Explain the meaning of the colors with a legend at the bottom of the chart.

- First consider livestock events such as bulling (red), calving or lambing (blue), weaning (yellow).

- Add wildlife and other considerations such as hunting season. Use the brainstorming list—and think.

Assess the Paddocks One by One

Cross out, with a line, any time periods when a paddock will be definitely unavailable to livestock—because of, say, recent fire, lack of water, heavy snow, hay cutting, field preparation for crops, or game management factors. Color coding or written notes will add clarity.

In row 26 under each month note the number of paddocks available. (Count as available any paddocks closed for less than half the month or less than two-thirds the longest recovery period; you can probably schedule grazing around the downtime.)

1 PREVIOUS NONGROWTH ADA YIELD / ADA RATING	2 ESTIMATED RELATIVE PADDOCK QUALITY	3 PADDOCKS NO.	SIZE	JANUARY	FEBRUARY	MARCH	APRIL	MAY	JUN
				SNOW COVER			CALVING		
		A	2000						
		B₁	850						
		B₂	870						
		C	2200						
		D₁	560						
		D₂	250						
		D₃	370						
		D₄	320						
		E₁	2500						
		E₂	400						
		F₁	750						
		F₂	1000						
		H₁	200						
		H₂	450						
		H₃	400						

25. NUMBER/SIZE OF HERDS	1/800 COWS & HEIFERS	1/800 COWS & HEIFERS	1/800 COWS & HEIFERS	1/800 COWS & HEIFERS	1/800 COWS & HEIFERS	1/800 C HE
26. PADDOCKS AVAILABLE		10	15	15	15	15
27.						
36. CELL SIZE	13,120 ACRES STOCKING RATE GROWING SEASON: 1:16.4			STOCKING RATE N		

AV. ANNUAL PRECIPITATION: _____

SEASON TOTAL PRECIPITATION: _____

REMARKS:

Calving	—··—··—··—	Poisonous
Breeding	— — — — — —	Leave gras
Weaning	✖✖✖✖✖✖✖✖✖✖✖	Graze to s
Hunting	— —·— —·— —	Snow cover
Elk Migration	▨▨▨	Haying

Note Special Management Needs

Clearly mark (again with coded colors) each paddock that has special problems or limitations—proximity to crops or a neighbor's bulls, poisonous plants, problem seedheads, lack of shelter, and so on. Don't forget to consider also any multiple use and wildlife factors and parasite cycles that can be broken with careful planning. (See page 65 in "Mastering the Basics" for detailed informa- tion.) Now you are ready to mark paddocks that require special treatment for management reasons. There may be a fire threat you want to graze off early in the dormant season, a brush area where you need to either advance or retard succession, or a cropland that should be either grazed or trampled. This is where you should plan to use such tools as grazing and animal impact to achieve your landscape goals. Otherwise you'll never reach them.

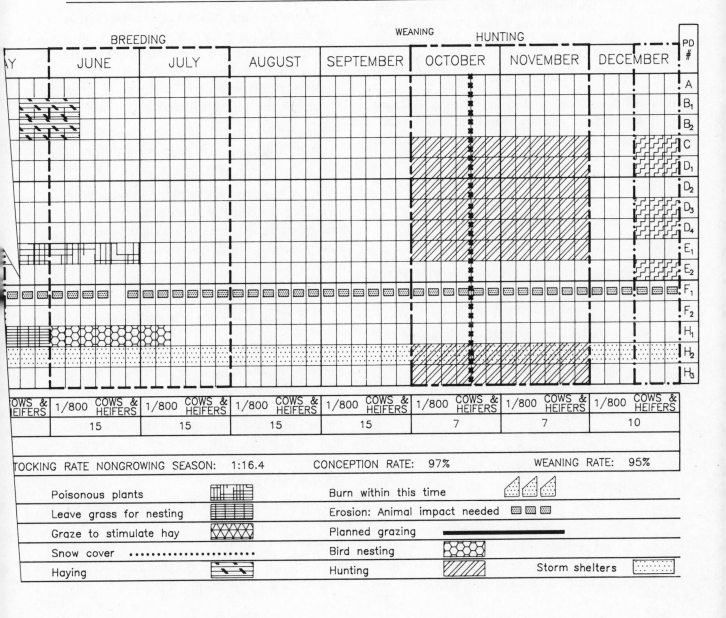

Poisonous plants	Burn within this time
Leave grass for nesting	Erosion: Animal impact needed
Graze to stimulate hay	Planned grazing
Snow cover	Bird nesting
Haying	Hunting	Storm shelters

Step Three:
Handling Livestock in the Cell

First decide if animals will run as one herd for the whole season—or, if not, how many you'll have and during what periods. Page 62 in "Mastering the Basics" gives several alternatives:

• Allow herds to use any paddock.

• Assign certain paddocks to each herd (dividing the cell into smaller cells).

• Use follow-through grazing where the second herd enters a paddock as the first leaves. Though this plan may offer the best solution to situations, it's tricky and requires constant monitoring.

Then enter on line 25 the number and size of each herd in each month. If you're planning a follow-through grazing pattern, note which herd goes first.

Step Four:
Deciding the Range of Recovery Periods

During seasons of active growth remember:

• Shorter and longer recovery periods reflect the length of time that severely bitten plants need to recover.

• The faster the growth, the shorter the recovery period; the slower the growth, the longer the recovery period.

• On arid and semiarid land, 30 to 90 days will usually suffice. Given higher or more effective precipitation and more fibrous vegetation, 20 to 40 or 60 days may do. For pastures (irrigated or not), particularly those with runner-type grasses, try 15 to 30 days.

• These times are only guidelines. In prolonged periods of adverse growth you may have to lengthen recovery periods further. (See page 60 on amalgamating herds to extend recovery periods and increase stock density.)

During nongrowing months remember:

• Recovery periods will not affect plant health much, but the timing is important for cropland soils, livestock performance, wildlife, and management factors such as weather protection, water, and access.

• A grazing strategy that rations out standing forage efficiently for both stock and wildlife will cut supplement needs. Generally this means slowing down the moves so that stock will graze each paddock only once and won't take all the best forage at the beginning of the season. (See the advice on nutrition on page 57 or for more detail consult the HRM textbook.)

• Your strategy must include a "time reserve" for drought and a margin for wildlife.

Now record your maximum and minimum recovery periods for the planned season on row 27 of the chart. Always indicate a range for growing seasons. If a single time will describe your dormant-season strategy, just record that.

Step Five:
Rating Paddock Productivity

Estimate Forage Quality

Rate the forage quality of each paddock in ADA and enter the number under the slash in column 1 of the chart. Remember: The figure is for comparison of quality only and may or may not reflect the actual growing-season yield. You can make this rating in several ways:

• Adjust the nongrowing season ADA from a season previously recorded above the slash in column 1.

• Whatever the season, the method described in "Mastering the Basics" for estimating forage reserves will show what each paddock could yield during one grazing. In growing seasons this figure will not give you total productivity because the paddock may be grazed several times, but it will indicate relative quality. In each paddock, pace off a square that could feed one animal for a day and compute the area in square yards. Divide into the number of square yards in one acre (4,840) to get ADA.

• Use past growing season records to estimate the ADA yield for one grazing.

Note that you can also grade forage quality, like schoolwork, by percent. An irrigated field, for example, might rate 140% and a barren upland 60%. Use this method if you can't get an ADA figure from either records or monitoring—but switch to ADA as soon as possible.

Rating Relative Paddock Quality

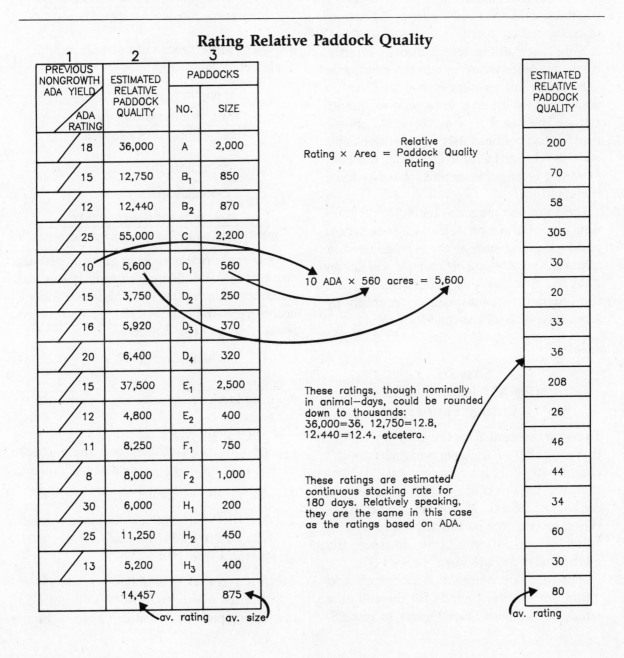

1 PREVIOUS NONGROWTH ADA YIELD / ADA RATING	2 ESTIMATED RELATIVE PADDOCK QUALITY	3 PADDOCKS NO.	3 PADDOCKS SIZE
18	36,000	A	2,000
15	12,750	B_1	850
12	12,440	B_2	870
25	55,000	C	2,200
10	5,600	D_1	560
15	3,750	D_2	250
16	5,920	D_3	370
20	6,400	D_4	320
15	37,500	E_1	2,500
12	4,800	E_2	400
11	8,250	F_1	750
8	8,000	F_2	1,000
30	6,000	H_1	200
25	11,250	H_2	450
13	5,200	H_3	400
	14,457		875

av. rating   av. size

$$\text{Rating} \times \text{Area} = \frac{\text{Relative}}{\text{Paddock Quality}}\ \text{Rating}$$

10 ADA × 560 acres = 5,600

These ratings, though nominally in animal—days, could be rounded down to thousands: 36,000=36, 12,750=12.8, 12,440=12.4, etcetera.

These ratings are estimated continuous stocking rate for 180 days. Relatively speaking, they are the same in this case as the ratings based on ADA.

ESTIMATED RELATIVE PADDOCK QUALITY
200
70
58
305
30
20
33
36
208
26
46
44
34
60
30
80

av. rating

Relate Forage Quality and Paddock Size

Multiply the paddock ADA rating by the size in acres and enter the result in column 2 ("Estimated Relative Paddock Quality"). These numbers, nominally animal-days, may be large, but you can list them in thousands for simplicity. Again, they have no meaning except to indicate the productivity of paddocks in relation to each other.

Now compute the average rating for all the paddocks and enter this figure at the bottom of column 2.

Note that you can rate paddocks directly, if you absolutely can't grade forage quality. If forage is fairly uniform, paddock size in acres will itself serve as a relative quality rating. The number of animals you believe could graze continuously in a paddock also reflects relative quality. The ADA method, however, is far superior and should become the standard.

The computer programs for biological planning ask for either an ADA rating or a whole paddock rating such as the acreage itself or a continuous stocking estimate. If you use an ADA figure or any other forage quality grade, the program will automatically multiply by acres to give a relative paddock rating.

Step Six
Calculating Maximum and Minimum Grazing Periods

This step is essential for the growing season. If the number of available paddocks doesn't change, you only have to do it once. Whenever any paddocks are out for periods you can't schedule around, you must do it again. Whether you rate paddock productivity in animals-days or some other measure, the mathematics are the same.

First calculate the *average* maximum and minimum grazing periods for the cell as a whole, and record these figures in row 28:

1. One herd in the cell:

$$\text{Av. max. GP} = \frac{\text{max. recovery period}}{\text{number of paddocks} - 1}$$

$$\text{Av. min. GP} = \frac{\text{min. recovery period}}{\text{number of paddocks} - 1}$$

2. Two or more herds, any paddock open:

$$\text{Av. max. (min.) GP} =$$

$$\frac{\text{max. (min.) recovery period}}{(\text{number of paddocks} \div \text{number of herds}) - 1}$$

3. Two or more herds with certain paddocks allocated to each:

$$\text{Av. max. (min) GP} =$$

$$\frac{\text{max. (min.) recovery period}}{\text{number of paddocks allocated} - 1}$$

4. Herds on follow-through grazing:

$$\text{Av. max. (min.) GP} =$$

$$\frac{\text{max. (min.) recovery period}}{\text{number of paddocks} - \text{number of herds}}$$

Now convert the average grazing periods you just computed into *actual* minimum and maximum grazing periods for each paddock according to the paddock ratings, and record these figures in column 4:

$$\text{Max. (min.) GP} =$$

$$\frac{\text{paddock rating}}{\text{av. paddock rating} \times \text{av. max. (min.) GP}}$$

If some paddocks have much longer grazing periods than others, you must check to see that recovery periods are adequate:

1. Add all the *minimum* GPs together.

2. From the total subtract the longest minimum GPs to find the *actual* recovery periods for these paddocks.

3. If any recovery period is much too short, you must add days to the minimum GPs in other paddocks that can absorb them.

4. Follow the same procedure for maximum GPs, though the problems will probably be less critical if you can't make complete adjustments.

Step Seven:
Planning the Grazing for the Growing Season

Using a soft pencil, mark the grazings in each paddock, showing the length of time by the length of the line. Remember:

- Use maximum grazing times in planning for greater safety. (Remember that the greatest overgrazing danger comes from moving too fast during slow growth.) When your monitoring shows rapid growth, shorten the grazing periods in practice.

EXAMPLE
Average Grazing Periods for Multiple Herds

Given: 3 herds, 75 paddocks (any herd can use any paddock), 30-90 day recovery periods

$$\text{Av. max. GP} = \frac{90}{(75 \div 3) - 1} = 3.75$$

Given: 8 paddocks assigned to herd A, 12 paddocks assigned to herd B

$$\text{Av. max. GP herd A} = \frac{90}{8 - 1} = 12.9$$

$$\text{Av. max. GP herd B} = \frac{90}{12 - 1} = 8.2$$

Given: 2 herds on follow-through grazing, 20 paddocks, 30-90 day recovery periods

$$\text{Av. max. GP} = \frac{90}{20 - 2} = 5$$

Figuring Grazing Period Guidelines

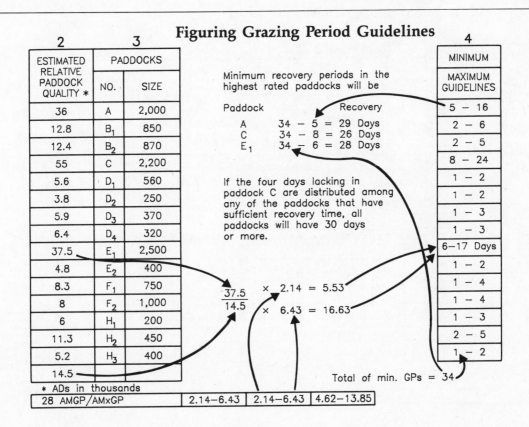

ESTIMATED RELATIVE PADDOCK QUALITY *	PADDOCKS NO.	PADDOCKS SIZE		MINIMUM MAXIMUM GUIDELINES
36	A	2,000		5 — 16
12.8	B₁	850		2 — 6
12.4	B₂	870		2 — 5
55	C	2,200		8 — 24
5.6	D₁	560		1 — 2
3.8	D₂	250		1 — 2
5.9	D₃	370		1 — 3
6.4	D₄	320		1 — 3
37.5	E₁	2,500		6—17 Days
4.8	E₂	400		1 — 2
8.3	F₁	750		1 — 4
8	F₂	1,000		1 — 4
6	H₁	200		1 — 3
11.3	H₂	450		2 — 5
5.2	H₃	400		1 — 2
14.5				

Minimum recovery periods in the highest rated paddocks will be

Paddock	Recovery
A	34 — 5 = 29 Days
C	34 — 8 = 26 Days
E₁	34 — 6 = 28 Days

If the four days lacking in paddock C are distributed among any of the paddocks that have sufficient recovery time, all paddocks will have 30 days or more.

$$\frac{37.5}{14.5}$$

× 2.14 = 5.53

× 6.43 = 16.63

Total of min. GPs = 34

* ADs in thousands

28 AMGP/AMxGP	2.14—6.43	2.14—6.43	4.62—13.85

- Allow for special conditions marked earlier—such as poison plants, wildlife, or a preferred calving ground. In paddocks with a poison plant danger, for instance, you might cut grazing times to the minimum so that grazing pressure won't force consumption of the problem plants.

- The graphic layout of the planning chart makes it possible to mark these obligatory grazings in first and then plan other paddocks forward or backward in time. This can mean staying longer in a paddock or two or moving across the cell to get the best effects at critical times. With a routine rotation schedule that you can't see on paper, you lose this advantage.

- Check to make sure recovery times are sufficient. If you've shortened the planned grazing for special cases, you may have to add extra days to grazing periods in other paddocks that can bear the extra time.

Now compute the monthly animal unit figures for each month and enter them in row 29. Rows 30 to 35 will help you keep track of different classes of livestock. Use the tables in "Mastering the Basics" to convert to standard animal units (SAUs) for a more accurate picture. For a better picture of the demands on your land you might compute the ADA or SDA taken at each grazing and note it by the grazing line on the chart.

Step Eight:
Computing the Stocking Rate
for the Nongrowing Season

First estimate the quality of each paddock in ADA and enter the figure in column 5. If the

Planning Chart with Grazings Marked In

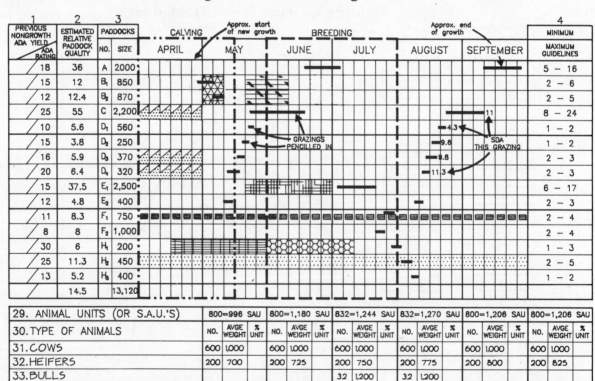

29. ANIMAL UNITS (OR S.A.U.'S)	800=996 SAU			800=1,180 SAU			832=1,244 SAU			832=1,270 SAU			800=1,206 SAU			800=1,206 SAU		
30. TYPE OF ANIMALS	NO.	AVGE WEIGHT	% UNIT	NO.	AVGE WEIGHT	% UNIT	NO.	AVGE WEIGHT	% UNIT	NO.	AVGE WEIGHT	% UNIT	NO.	AVGE WEIGHT	% UNIT	NO.	AVGE WEIGHT	% UNIT
31. COWS	600	1,000		600	1,000		600	1,000		600	1,000		600	1,000		600	1,000	
32. HEIFERS	200	700		200	725		200	750		200	775		200	800		200	825	
33. BULLS							32	1,200		32	1,200							

information is available, use the dormant-season yield from past seasons (column 1) as a guide—but adjust for weather, regrowth since last grazing, and so forth. Multiply by paddock area (column 3) to get the estimated animal-days. Enter that figure in column 6 and enter the total in row A.

Then enter the expected days of dormancy in row B. If you expect prolonged snow cover, subtract that from the row B figure and note what you did. Work out the feed requirement plus a reserve for late blizzards and enter that on line 24 ("Supplement Type and Amount").

On row C, enter a "time reserve" for drought or late return of growth.

Enter the total of rows B and C on row D.

Divide the animal-days in row A by days in row D to get the estimated carrying capacity for row E.

Step Nine
Planning the Grazing for the Nongrowing Season

Mark the grazings, paddock by paddock, with a soft pencil. Here you will implement a strategy for ease in management and good nutrition. (See page 57 in "Mastering the Basics.")

Using the animal unit figures in row 29, figure the ADs (or SDs) you plan to take from each paddock and enter the number in column 7. If you are worried about your animals

Nongrowing Season Plan

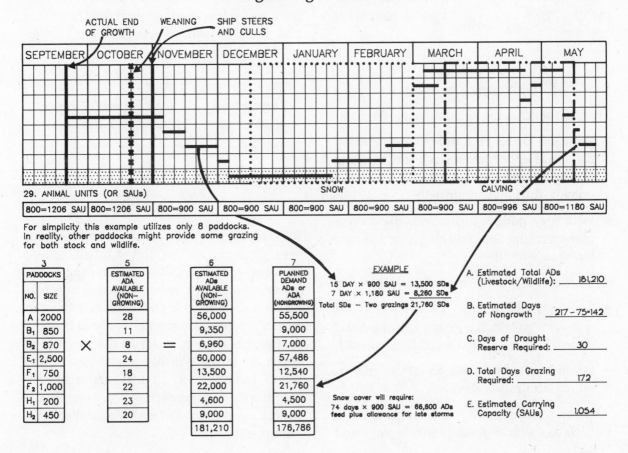

29. ANIMAL UNITS (OR SAUs)

| 800=1206 SAU | 800=1206 SAU | 800=900 SAU | 800=900 SAU | 800=900 SAU | 800=900 SAU | 800=900 SAU | 800=996 SAU | 800=1180 SAU |

For simplicity this example utilizes only 8 paddocks. In reality, other paddocks might provide some grazing for both stock and wildlife.

3 PADDOCKS		5 ESTIMATED ADA AVAILABLE (NON-GROWING)	6 ESTIMATED ADs AVAILABLE (NON-GROWING)	7 PLANNED DEMAND ADs or ADA (NONGROWING)
NO.	SIZE			
A	2000	28	56,000	55,500
B₁	850	11	9,350	9,000
B₂	870	8	6,960	7,000
E₁	2,500	24	60,000	57,486
F₁	750	18	13,500	12,540
F₂	1,000	22	22,000	21,760
H₁	200	23	4,600	4,500
H₂	450	20	9,000	9,000
			181,210	176,786

With × between columns 3 and 5, and = between columns 5 and 6.

EXAMPLE

15 DAY × 900 SAU = 13,500 SDs
7 DAY × 1,180 SAU = 8,260 SDs
Total SDs — Two grazings 21,760 SDs

Snow cover will require:
74 days × 900 SAU = 66,600 ADs
feed plus allowance for late storms

A. Estimated Total ADs (Livestock/Wildlife): ___181,210___

B. Estimated Days of Nongrowth ___217 - 75 = 142___

C. Days of Drought Reserve Required: ___30___

D. Total Days Grazing Required: ___172___

E. Estimated Carrying Capacity (SAUs) ___1,054___

Actual Grazing Record, Recorded on the Chart

YEAR 19 _____

1 PREVIOUS NON GROWTH ADA YIELD / ADA RATING	2 ESTIMATED RELATIVE PADDOCK QUALITY	3 PADDOCKS NO.	SIZE	JANUARY	FEBRUARY	MARCH	APRIL	MAY	JUNE	JULY
				SNOW COVER			CALVING	GROWTH STARTS	BREEDING	
13.5 / 8	8	F₂	1,000		14,400 SD FED					5
22.5 / 30	6	H₁	200		18,000 SD FED					
20 / 25	11.3	H₂	450	35,000 SD FED						

	JANUARY	FEBRUARY	MARCH	APRIL	MAY	JUNE	JULY
21. RAINFALL			2 1		2 2 1½ 1		2 3
22. SNOW	14 8	5		8			
23. GROWTH RATE (F/S/O)							
29. ANIMAL UNITS (OR SAUs)	800=900 SAU	800=900 SAU	800=900 SAU	800=996 SAU	800=1,180 SAU	832=1,244 SAU	832=1,270 SAU

The piece of the grazing plan here shows how the record of a cell looks after the year is over. The actual grazings (often different from the planned ones) are inked in. The yield (in SDA or ADA) for each grazing in each paddock is written after each grazing line.

When cattle are fed during snow cover, the number of stock-days fed is written in instead. The total yield for the dormant season (in ADA or SDA) is written in column 1 to the left of the slash where it can serve as the basis for an ADA rating written to the right of the slash.

and/or wildlife, check in the field, using sample grazing areas needed by one animal for one day.

Step Ten:
Operating the Plan

During the growing season, remember:

- With low paddock numbers, overgrazing can occur either when stock stay too long during fast growth *or* when they return too quickly during slow growth.

- With high paddock numbers, the danger of overgrazing is in returning too soon during slow growth.

- During fast growth, move livestock faster.

- During slow growth, move livestock as slowly as nutritional needs and the Maximum Grazing Period guideline permit.

- If you have any doubts about the growth rate, assume it is slow.

In nongrowing seasons, use longer grazing periods as much as possible so that stock and

wildlife do not take the best out of every paddock early, leaving the worst for the critical end of the season. (See page 57 in "Mastering the Basics.")

Thoughtless rotation is a major cause of stock stress in the dormant season.

Step Eleven:
Keeping the Record

To fine-tune your yield estimates and management decisions in the future, record actual events in ink as the season progresses. For every paddock, ink in a line on the chart for the time a herd spent there. Herders can record this figure in the field for you by daily dropping a stone in a can on the gatepost of paddocks in use. Washers on a nail will also serve. Indicate with letters—for light (L), medium (M), or heavy (H)—how severely the paddock was grazed when the stock left it (for example, 10/M).

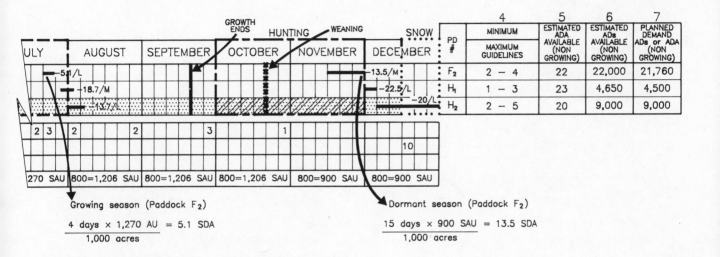

Growing season (Paddock F₂)

4 days × 1,270 AU = 5.1 SDA
1,000 acres

Dormant season (Paddock F₂)

15 days × 900 SAU = 13.5 SDA
1,000 acres

Record precipitation on rows 21 and 22. The blanks represent 5-day periods. You may want to note the exact day of major storms with a dot and a written comment elsewhere on the chart.

Put your opinion of daily growth rates in the same 5-day periods on row 23 using letters for slow (S), fast (F), or no (0) growth.

Shortly after the growing season has begun, make your best guess of an actual starting date and put an indelible green line down through all paddocks for that day. When the season has definitely ended, do the same thing to indicate the beginning of the dormant season.

If you didn't convert to standard animal units during the planning, consider doing so now. (See page 43 in "Mastering the Basics.")

For each inked-in grazing, figure the ADA taken and record it next to the line. Then:

• For each paddock, add the ADA taken during the *dormant season* and enter the total to the left of the slash in column 1. This will help you make an ADA estimate for future seasons. (In these figures, do not count feed supplied.)

• Use the ADA figures for the growing season to help in reassessing the relative paddock ratings in column 2, recovery periods, and the like. The actual performance of livestock in a paddock will tell you more about forage quality than any known measuring technique. Do not total the growing-season ADA. Individually they tell you a lot about yield *per grazing* and adequacy of recovery periods. A total only tempts you to think you could harvest that much in a single grazing.

Add any comments that will guide you in future planning or management decisions. The back of the chart provides all the space you need.

At year's end record in the lower right corner the livestock production results—particularly the yield per acre, as production per unit of land usually means more than yield per animal.

Summary Livestock Results

Calving/lambing/kidding	96%
Av. weaner weight age	565/6 MTHS.
Daily av. weight gains	1.0 LBS.
Total yield per acre	30.85 LBS.

Gain on weaned calves:
576 head × 565 lbs = 331,776 lbs

Gain on retained heifers:
200 head × 365 lbs = 73,000 lbs
Total   404,776 lbs

404,776 lbs ÷ 13,120 acres = 30.85 lbs/acre

Guidelines for Simplified Grazing Management

The planning chart described in the preceding section allows you to handle many variables while assuring maximum animal performance, but only a fairly disciplined and diligent person can fill out the chart and put it to work. If you can't plan with the *Aide Memoire*, there is a simpler process that can still help your land. While these guidelines may not provide maximum livestock performance or achieve highly refined landscape goals, they will help you to minimize overgrazing and increase your stocking rate.

In practice a rancher, adviser, or someone who can handle a bit of paperwork uses these guidelines to set up maximum and minimum grazing times for the grazing areas available (fenced or unfenced). Then as long as the operator stays within that framework and carries out simple monitoring to avoid overgrazing or overstocking, all should go well.

Step One:
Deciding the Range of Recovery Periods

First decide the range of recovery periods for the season in question. (This is essentially Step Four in the *Aide Memoire*.)

During seasons of active growth remember:

- The shorter and longer recovery periods reflect the length of time that severely bitten plants need to recover in fast-growth and slow-growth conditions.

- On arid and semiarid ranges, 30- to 90-day recovery periods usually suffice. Given higher or more effective precipitation and more fibrous vegetation, 20 to 40 or 60 days may do. For pastures (irrigated or not), particularly those with runner-type grasses, try 15 to 30 days.

- These times are only guidelines. In prolonged periods of adverse growth you may have to lengthen recovery periods further. (See also page 60 on amalgamating herds to extend recovery periods and increase stock density.)

During nongrowing months remember:

- Recovery periods are not a major factor in plant health, but timing may be important to cropland soils, livestock performance, wildlife, and management factors such as weather protection, water, and human access.

- A grazing strategy that rations out forage efficiently will cut supplement needs and help wildlife. Generally you'll want to slow down moves so that stock will graze each paddock only once and not take all the cell's best forage early in the season. (See the advice on nutrition on page 57.)

- Whatever your grazing strategy, it must include a "time reserve" and a margin for wildlife.

Recording grazing days in a paddock.

Step Two:
Figuring the Grazing Periods

Work out maximum and minimum grazing periods for all paddocks or grazing areas in the cell. Follow the same procedures described in Steps Five and Six of the *Aide Memoire*. But instead of putting the numbers on a planning chart, post them on paddock gates or boundaries or a simple map on the bunkhouse wall. If the people responsible for the herd move animals according to these guidelines, they'll avoid most problems associated with overgrazing.

Step Three:
Establishing Day-to-Day Procedures

First set up a *simple* system for recording how long a herd spends in one paddock. Stones in a can, washers on a nail, notches in a stick, or marks on a map will do, but the accounting should occupy the place where the operator customarily makes the decision to move and where grazing guidelines are displayed. No matter how simple, this is the hardest routine to maintain. People tend either to fall into a pattern of standard moves or to decide on moves by the state of the paddock occupied at the moment, often upsetting far more critical recovery times.

During the growing season:

• The operators should use the maximum grazing period guideline, because most growth is slow and there is less danger of making serious mistakes. Extend this period only with good reason, however, or animal performance and plant recovery may suffer.

• As stock leave each paddock, the operators should mark heavily grazed plants so they can monitor the rate and degree of regrowth. This tactic will alert them to periods of rapid growth and prevent them from keeping livestock in paddocks too long. In the absence of a formal plan that predicts the sequence of paddocks, you should avoid falling into a habitual cycle that ignores changes or opportunities.

• On clear proof of fast daily growth, operators should use the minimum guideline or close to it—particularly if you have few paddocks and thus long grazing periods. When growth slows, return to the longer guideline or close to it. Moving too fast during slow growth causes extreme overgrazing problems when paddock numbers and stock densities are high.

• Take any overgrazed or overbrowsed plants seriously. If necessary, get help in determining the cause. Usually it involves failure to time moves to the prevailing growth rate.

• Don't move animals according to a fixed pattern. Not only does this suppress creative thought but any routine cycle compounds mistakes and imbalances. Radial cells in which all paddocks share a common point make it easy to rotate herds like clock hands (an unfortunate temptation). If moving animals to an adjacent paddock, go directly through a seldom used gate—not through the center where cattle habitually trail.

Remember that in the dormant season you don't have to follow grazing-period guidelines. Lengthen or shorten them for management reasons—reducing the fire risk, giving stock storm shelter, evening out grazing, and so on. Keep stock moving to avoid the fouling produced by a concentrated herd. But slow the herd as described above (and on page 57) so that stock and wildlife can enjoy a first selection from some paddocks late in the season if you have enough paddocks. Remember: It's hard to achieve an even plane of nutrition with only a few paddocks.

Step Four:
Checking Stocking Rates

Contrary to popular belief, overstocking is rarely the primary cause of range problems. Regardless of range condition and even without the higher level of planning, planned grazing usually allows you to exceed—even double—conventional or government-prescribed stocking rates. Increased animal impact only benefits the land, though the margin left for wildlife may be cut if you don't plan fully with the *Aide Memoire*.

First consider other explanations for poor animal performance:

• Stocking rate too low, causing loss of forage value due to unharvested old growth or overrank plants

• Similar problems due to low stock density leading to high proportions of overrested plants

• Bad judgment on paddock moves at either low or high stock densities leading to high proportions of overrested plants

Look for overstocking by these signs:

• During the growing season—serious depletion of paddocks even during rapid growth and short grazing periods

• During the dormant season—ADA estimates (see page 52) that show inadequate standing forage to carry animals until growth starts again

If animals come to the point of reducing soil cover by eating litter, they're hurting the land.. You will have to lower the pressure by stock reduction, herd amalgamation, or, if some paddocks appear unaffected, by shifting the pressure to *them*.

Step Five:
Reassessing Paddocks

Your original judgment of paddock productivity will probably not be right, even if you know your land well. Weather, higher stock densities, and planned grazing may cause big changes in one season. Pay special attention during the dormant season if some paddocks run out of forage sooner than expected. Correct the problem by adjusting your paddock assessments and grazing guidelines.

Step Six:
Reviewing the Limits of the Simplified Guidelines

These simplified guidelines will help you limit overgrazing and broadly affect succession. Within this framework you can easily use such techniques as amalgamation of herds to increase density and extend recovery periods during times of drought, to stimulate herd effect through supplemental feeding, and so on.

Since it only handles grazing and recovery time in relation to plant growth, the simplified procedure does not serve in these cases:

- Where there is more than one herd per cell. Two or more herds, especially when part of a follow-through grazing strategy, call for the full written plan and the *Aide Memoire*.

- Where the livestock policy must fit around other considerations.

- Where wildlife habits or habitat call for special treatment or there are many other simultaneous uses of the land.

- Where maximum animal performance depends on factoring in breeding seasons, poisonous plant dangers, markets, and the like.

SUMMARY

The *Aide Memoire* for Biological Planning and the concepts behind it represent 30 years of trial and error on three continents in many situations. It works. Each step builds on the next one, so the order is important. Don't skip any. If they truly don't apply to your situation, pass on—but think about them anyway. You'll find yourself in control of an amazing array of varied and subtle factors affecting your operation.

There must be something philosophically significant in the fact that of all the procedures in this book the one for biological planning is the most orderly. Its object—the health and growth of plants and animals—is of course the most unpredictable, complex, and baffling subject that human intelligence ever sought to fathom. Without thinking this paradox all the way through, you still have to live with its significance and realize that those who don't recognize it fail.

Because living organisms never stop changing and even microbes have plans of their own, you yourself can never stop planning. If you do, you will fall into habits and routines that worked when you first planned them but will make you miss the opportunities of the ever-changing game. Eventually they will lead to ruin.

The human weakness for routine runs deep, but we've come a long way from our wild roots, and we often confuse the wonderfully dynamic patterns of nature and the reassuring regularity of machines. The greening cottonwoods in spring, the wedges of geese in fall, the grasshoppers of summer, as timeless in their regularity as they may seem, are not the same as the New York to Los Angeles Boeing overhead.

You plan for living things, not because you can ever hope to bind them to your pattern, but so you can fit yourself into theirs. Thus you must do it every year and sometimes more often to stay with the game.

Perhaps the routines for holistic biological planning must be so ordered and strict because of nature's diversity and complexity. Of all the challenges of management, we are culturally and mentally least equipped to understand this one. We must proceed like a tightrope walker with a death grip on the balance bar.

Nothing is perfect, though, and the *Aide Memoire* for Biological Planning that has evolved continuously over the years will continue to develop. Do your planning, keep track of your ideas, and if you develop improvements, let us know.

PART III
BIOLOGICAL MONITORING

BIOLOGICAL MONITORING

A fundamental axiom of holistic management says that success requires planning—and successful planning takes monitoring, controlling, and replanning. The textbook makes this point in many ways and contexts, and earlier chapters in this workbook have repeated it. Nevertheless, monitoring developments in the biological sphere deserves its own treatment, because most people unfamiliar with holistic management have some strong preconceptions to overcome.

The livestock industry traditionally does monitor many aspects of animal performance. In the cattle business, the statistics on conception rates, bull performance, daily gain, weight per day of age, calf and weaner weights, more or less define the quality of an operation in traditional terms.

The holistic manager generally cares more about land performance, as expressed in yield and cost per acre, than animal performance. Thus really good monitoring should focus on factors that will expose problems in this area before they affect results.

The old-time coal miners, for example, found it wiser to monitor air quality in the shaft by counting dead canaries instead of dead miners, because canaries died first. Similarly, a drop in conception rates does show a problem—but after the fact and without any clue of how to correct it. Many of the numbers we ardently compile and ponder fail us in the same way.

Obviously you can steer a ship better looking over the bow than back at the wake, but only if you know what to monitor. Ideally biological monitoring should pick up changing conditions and deviations from plan before conception rates fall or daily gain drops off, so you won't miss an opportunity to change course and replan.

You must address this challenge on three levels. First you must cultivate a general and ongoing awareness of the conditions of the four ecosystem foundation blocks (succession, water cycle, mineral cycle, and energy flow) and how your tools affect them.

Second, you must periodically carry out a detailed assessment of the soil surface and the life upon it. The procedure differs fundamentally from a traditional forage inventory because it notes factors that foretell changes and trends rather than the present feed production.

Third, you must monitor such daily developments as grazing and animal impact, growth rates of plants, and water supplies. These factors are the meat and potatoes of daily management decision-making, but other indicators are equally important because of their long-term implications.

These three levels of monitoring complement each other. When practiced together they will quickly reveal the dynamism present in all landscapes. Most people who make a living from the land or love it deeply find this revelation fascinating. The resulting habit of observation becomes addictive but hard to communicate to people who haven't tried it. A Colorado rancher once confessed to lying to his guests to avoid the trouble. "At chore time I tell them I'm going to check the cattle," he said. "They wouldn't believe that I'm really checking the grass."

Like the preceding parts of the book, this one presents some basic concepts and examples that illustrate the holistic approach to biological monitoring and then a step-by-step guide to the formal procedure. The advice, of course, is not exhaustive, as the possibilities have no end. They nevertheless provide a starting place. And if you haven't looked at land this way before, no land will ever appear the same again.

MASTERING THE BASICS

General Observations

As mentioned earlier, good monitoring depends on a broad awareness of the state of the land, but this means training yourself to look for a broad range of specific things. If you happen to be a hunter, you know that putting your mind on deer tends to program your eyes to see deer. Tomorrow you might walk the same woods hunting rabbits or turkeys and not notice the buck you would have shot yesterday.

So with monitoring, you must consciously direct your thoughts in order to see. And don't just think about the ground. Look in books and records, in the expertise of others, and in your own experience.

The monitoring described here has been developed for rangelands and watersheds, but parallel methods should be developed for croplands, complex forests, and wildlife management areas where monitoring is essential.

Because arable land usually does not approach rangeland in size or variety of topography and plant life, farmers may well have less difficulty in perfecting simple procedures for monitoring soil cover, aeration, structure, quality and quantity of organic matter, and abundance of living organisms.

Historical Data

Monitoring has no meaning except in relation to your goals and your plan to reach them. Progress toward a goal, however, implies a starting point—and your monitoring will be much more perceptive and useful if you know where that starting point lies in relation to the past use and potential use of your land.

If you've read, for instance, that the U.S. Cavalry in 1850 harvested its winter hay from what has since given way to desert, this knowledge will allow you to assess the present level of succession correctly and alert you to vestiges of lost plant communities that your plans may revive. If you've just purchased a tree farm in the Carolinas and do not know that it once supported a prosperous tobacco plantation, you might suppose that it could never grow anything but loblolly pine and therefore not look for improvement.

When you begin managing a piece of land, photograph the characteristic sites, take notes of their condition, and try to reconstruct what has happened there in the past. Useful information might include the location of springs and streams that have dried up or become intermittent, the dates and extent of past fires and floods, and observations about changes in crops, wildlife, and plant communities. You'll want to back up these general observations with hard data from fixed-point photographs and the transects described at the end of this section—but, again, general observations are equally important. The form might be something like the following:

Sheep Creek Drainage, June 1986

According to county historians, Sheep Spring was an important watering point on sheep drives to the railhead at Upstart. Over 10,000 passed in the summer of 1879. Evidence in graffiti on rocks. The creek must have been quite different as accounts mention frequent fights between herders when bands mingled across a shallow stream that meandered through a meadow.

Tradition says Indians had an antelope trap in the canyon at south end of the valley. Arrowheads found there still.

In 1903 there was a range fire hot enough to destroy a prospector's cabin.

The Vee Bar, a huge open-range cattle operation, included the valley in its summer range until 1929. Intermittent use by squatters continued till 1940, when it was fenced. Continuous grazing continued until 1983 when it was acquired by Vesuvius Life Insurance Co. No stock or formal management until 1986.

After completing the transects of vegetation and soil surface condition described later, you might summarize the status of the four ecosystem blocks thus:

Succession:

Definitely lower than in the glory days. Mostly sage and annual grasses. Many young piñons. Anthills abundant. Sage, winterfat, cliff rose, and other shrubs show signs of severe past overbrowsing followed by regrowth that must have begun in 1983. Some winter use by deer evident. Isolated examples of perennial grasses (including sand dropseed, giant dropseed, Indian rice grass, western wheatgrass) often associated with yucca or less frequently with sage. Also some mats of blue gramma and a fair amount of spiny muhly.

Water Cycle

Generally ineffective and evidenced by a large percentage of bare ground. Most perennial plants on pedestals, often with exposed roots. Soil capping pervasive, often mature and black with algae. Many rills and small gullies. Most litter washed into heaps and banks, showing force of runoff. Sheep Creek is an intermittent arroyo with 6- to 10-foot vertical banks broken occasionally by old stock and game trails.

Mineral Cycle

Generally ineffective. Dung pats three years and older evident everywhere. Perennial grasses show accumulations of past growth. Annual grasses largely washed or blown away by midwinter. Deeper-rooted plants, sage, and piñon thriving.

Energy Flow

Obviously not high because of large amount of bare ground. Most sunlight energy that's converted to forage is not harvested by animals.

The Photo Record

Photographs show changes on the land better and more dramatically than any other record. Later we'll see how to take a scientific measurement of your land's health including several photos taken from the same spot each year. Aside from this rather time-consuming procedure, however, snapshots taken of the same scene in different seasons across a period of years broaden the record and may lead to important insights.

Carry a camera like many ranchers carry a fencing tool and a rifle. Get in the habit of shooting interesting anomalies, and get repeats of the same scene for comparison. In this way you can monitor the healing of a gully, the development of a fence-line contrast, animal impact around a water point or gate, the effect of a herd during one grazing period, the formation of trails, and any number of other concerns.

Succession

Level of succession is the key indicator of the land's stability, resilience, productivity, and health. Moreover, manipulation of succession is the principal way to achieve your landscape goals.

Assessing the level and direction of succession is a subjective business. Roughly speaking, succession is the sequence through which a biological community develops from a bare surface and is thus different for every climate and soil type. Human activity, fire, and other interventions distort the process in unique ways, and we seldom encounter a truly fresh start from bare ground anyway. Textbooks like to talk about "climax" communities as a definite endpoint of succession, but in reality few environments (virtually no brittle ones) enjoy stable conditions long enough to determine what this is. Thus you must judge the level of succession and its general direction by a variety of signs, some of which may conflict. Here are some of the main ones:

Simplicity vs. Complexity

This is one of the most important indicators. A wide diversity of species (rather than vast numbers of one species) indicates advanced succession. Even if total biological mass is impressive, monocultures of plants or vast numbers of one animal usually indicate a lower level of succession than the land can support. An exception might be the diversity of regrowth following a forest fire.

Seasonal diversity is also important. In most of North America an absence of either cool-season or warm-season grasses, for example, is a distortion of succession.

Annual vs. Perennial

A predominance of annual plants indicates a low level of succession. Among grass species, annuals generally grow a seedhead on every stalk. Perennial grasses have many stalks that produce only leaves and will probably show some sign of the previous year's growth. Perennial plants contribute far more than annuals to soil cover and stability through their greater root depth and organic production. They also give a more consistent picture of community health, because huge numbers of annuals may burst forth only when the right conditions coincide.

Presence of Certain Plants and Animals

Generally, complexity of species is a more reliable index of succession than mere numbers, but most species thrive only at certain successional levels. Low-level organisms tend to reproduce rapidly and depend on a simple environment. High production of seeds adapted to disperse and penetrate capped soil is typical of low-successional grasses. Many kinds of rodents and harvester ants are low-successional. Moose and fir trees are typically high-successional. Among domestic animals, dairy cattle have a limited niche high in succession, while sheep and goats span a broader range.

Status of Youngest Age Class

Trends in succession often show first in the presence or absence of young plants or animals. If the young of perennial grasses or high-successional animals do not survive, that foretells decline—as would extraordinary reproductive success of low-successional organisms. Likewise, young animals or plants that increase diversity or fill high-successional niches indicate advance. Among grasses, a drop in successional level often shows up as an increase in species that spread effectively by rhizomes or stolons rather than seeds.

There is no such thing as a hardy plant or animal. All organisms thrive at certain successional levels. Some, like humans and coyotes, range over a wide spectrum of successional levels. Others, like prairie dogs or flying squirrels, are confined. As succession advances, better mineral and water cycles will support a greater variety of species at all levels.

Presence of Woody Plant Species

Woody forbs, shrubs, and trees usually reflect an ascending order of succession—as in the case of a tropical cornfield returning to jungle. In brittle environments this matter is more complex, and woody species and trees often flourish in declining grassland. This is not an advance when it represents a loss of diversity and may be a passing phase as damaged water cycles eventually kill off the woody plants also.

Status of Ground Cover

Since capped, bare ground is by definition the bottom end of succession, the earliest sign of change is often an increase or decrease in the space between perennial plants—particularly grasses. Many organisms won't propagate without open friable soil. Areas too arid to sustain a cover of living plants require litter. Changes in litter and capping often precede changes in succession.

Vestiges of Old Communities

Many remnants of communities that thrived at higher levels of succession in the past may hang on in sites protected by cactus, rocks, and the like. They document the land's decline, indicate its potential, and provide the seeds for a comeback.

Economic Uses

When an area's livestock industry has to shift from cattle to sheep to goats or from production of meat to primarily hair, succession is generally declining. Conversely horses and dairy cattle indicate higher succession. In crop farming, a change in the kind of crops that can be grown or increasing need for chemical inputs also may reflect successional changes.

Water Cycle

Erosion

Erosion from wind and water obviously indicates a poor water cycle. Beneficial use of water simply does not include moving soil. Obvious gullies and rills indicate advanced erosion. More pervasive, however, and equally destructive is the sheet and rill erosion that leaves less obvious traces. Look for the following signs:

Pedestaling of Plants and Rocks

Sheet erosion, either by wind or water, will cut away bare soil, leaving roots up on little platforms. Occasionally the plant loses the battle and the soil is snatched from beneath it, leaving the roots exposed as stilts. In some cases plants actually accumulate soil, so the difference between the eroding surface and protected spots becomes extreme. In some parts of the American Southwest bushes like greasewood and Mormon tea now sit on hummocks 3 or 4 feet tall.

Plant pedestaling

Flow Patterns on Bare Ground

Moving water or wind scours the soil, leaving minute patterns. Multiplied by the area of land involved, the tonnage of soil carried off is substantial. Equally important, no organic matter can accumulate under such conditions.

Flow patterns

Litter Banks

Piles of litter jammed between plants and rock are a sign of more runoff of water than desirable, but the litter banks do stop some of it—showing once again the importance of litter.

Litter banks

Siltation in Low Points

The silt dropped behind check dams and in level areas below hillsides means excessive

and damaging runoff, even when it does not produce gullies.

Siltation

Splash Patterns

Raindrops, especially big ones that fall from tall trees, dislodge large particles of dirt when they hit bare ground. This action can move a surprising amount of soil and badly degrade the soil surface. Check the height of the mud splatters on plants and fence posts.

Dunes

Dunes are the end product of wind erosion. Even small dunes cause problems when they move and smother vegetation.

Permeability and Effective Rain

Ideally precipitation should enter the soil where living organisms can use it. This of course depends on ground cover, soil type and condition, aeration, organic content, slope, and other factors. If you know how different sites respond to rain, you'll be able to plan better use of sites and effective use of tools.

Long-term changes in soil permeability can cause dramatic changes in the water table and the disappearance or reappearance of springs. Hard, capped soil and lack of ground cover or litter indicate a problem with infiltration and aeration.

You can get a good idea of the situation by simply pouring a quart or so of water on the ground and timing how long it takes to disappear. If you wait 5 minutes and probe a bit with a shovel, you can also find out how effectively it spreads. A similar check after a short rain will give you some idea of how much growth response to expect from that rain.

Plant Habitat

Succession reflects the status of the water cycle. Plant types and associated animals will vary as soils range from dry to moist, from sealed to well-aerated, from well-drained to waterlogged, and as water tables rise or fall. Any changes in succession may indicate changes in the water cycle.

The textbook explains the relationship between plants and water cycle in some detail. A good look at your own land will probably reveal which species you can expect to increase and which will decrease as water cycles improve. In this context, remember that poor aeration, which occurs in both waterlogged and capped or compacted soils, rather than lack of water, may have the greatest effect on plants.

Mineral Cycle

The mineral cycle, like succession, has many aspects. Most of them, though, boil down to three questions you can check by observation:

• Are minerals visibly cycling?

• If not, what happens to them?

• How well are corrective measures working?

The health of the mineral cycle shows in the following ways:

A Simple Water Cycle Test

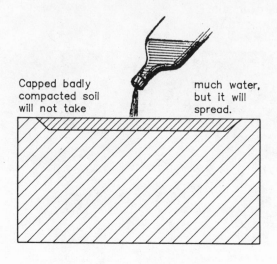

Capped badly compacted soil will not take much water, but it will spread.

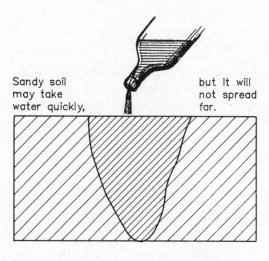

Sandy soil may take water quickly, but it will not spread far.

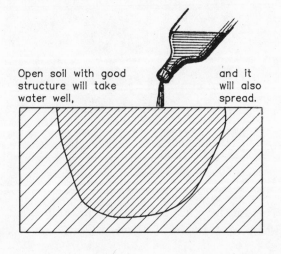

Open soil with good structure will take water well, and it will also spread.

Grass Types from Habitats That Reflect Water Cycle

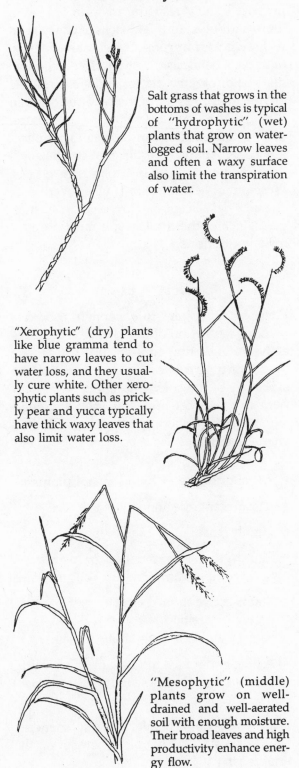

Salt grass that grows in the bottoms of washes is typical of "hydrophytic" (wet) plants that grow on water-logged soil. Narrow leaves and often a waxy surface also limit the transpiration of water.

"Xerophytic" (dry) plants like blue gramma tend to have narrow leaves to cut water loss, and they usually cure white. Other xerophytic plants such as prickly pear and yucca typically have thick waxy leaves that also limit water loss.

"Mesophytic" (middle) plants grow on well-drained and well-aerated soil with enough moisture. Their broad leaves and high productivity enhance energy flow.

Breakdown of Litter (Especially Dung)

A poor mineral cycle may allow dry dung pats to linger for years. Keep your eye on the amount of dead material that stays on plants and oxidizes, turning gray like old barn wood, instead of returning to the soil.

Activity of Soil Organisms

Many of these are microbial and can't be seen, but worms, ants, and burrowing rodents all enhance the mineral cycle. At certain successional levels some of these may boom to annoying numbers. The remedy often lies not in killing them back directly but in advancing succession to a degree of diversity that limits and stabilizes the population.

Presence of Plants of Varying Root Depth

Deep-rooted plants return leached minerals to the surface. In high-rainfall areas, loss of mineral-binding organic matter allows leaching to become extreme. Thus, farmers in wet climates must either apply extraordinary doses of fertilizer to "worn-out" pastures and croplands or let them revert to forest.

Livestock Consumption of Mineral Supplements

Your livestock's appetite for supplements often reflects missing elements in the natural mineral cycle. Livestock exposed to free-choice, cafeteria-style minerals will select whatever they don't get from grazing. In the case of some trace minerals, they may add enough to the natural cycle through their dung to eventually reduce need for supplements. As plant species concentrate minerals differently, a change in supplement consumption may also show the connection between the mineral cycle and successional diversity. (Free-choice mineral supplements have performed well in many livestock operations, but not in all. More research is necessary.)

Deficiency Symptoms in Plants and Animals

Poor mineral cycles may produce rough coats, infertility, or other weaknesses in animals. In crop plants leaves may yellow, curl, or develop purple veins. Soil tests tell part of the story; however, minerals may be present but unavailable or scarce yet concentrated by certain plants in certain places.

Soil pH and Sodium

Extreme acidity or alkalinity makes many nutrients unavailable to plants. Sodium destroys soil structure. More often than many like to admit, however, these problems also are greatly exacerbated by the lack of organic material that is characteristic of low or retreating succession.

Rather than countering problems by merely testing the soil and adding the missing ingredient, consider that succession might offer a better solution. Plants with varying root depths will recycle leached minerals. And a management plan that returns organic material to the soil will impede leaching and offset the effects of extreme acidity, alkalinity, and sodium.

For insight into the mineral cycle, monitor succession with an eye to:

- Diversity of plants, particularly nitrogen-fixing legumes

- Agents of decomposition, including the breakup of old plant material through animal impact

- The trend of organic material on and in the soil

- Varying root depths

Use your observations of all these factors in management decisions. The HRM model may well show that the manure spreader, more paddocks, or a few sacks of clover seed will serve your purpose better than chemical fertilizer. Chemicals tend to sterilize soil, causing loss of organic matter and soil-dwelling organisms (a retreat of succession). This in turn leads to breakdown of soil structure, accelerated leaching, and escalating application costs. Applying fertilizer in several small doses rather than one big one vastly reduces these complications. But whenever your chemical applications do pass the testing guidelines of the model, you must monitor for any ill effects.

Energy Flow

Energy flow is hard to observe in a precise way. It's a function of the total area of leaves actively converting sunlight into forage, the length of time this conversion goes on, the efficiency of the conversion, and what happens to the forage after it is grown. Generally speaking, then, high energy flow is characterized by:

- An abundance of broad-leafed plants

- Close plant spacing

- Highly productive plant types

- Rapid growth unimpeded by poor aeration, compaction, or other soil problems

- Plants active through the longest possible growing period (both warm and cool-season plants in temperate climates)

- A long and robust food chain—herbivores, predators (including commercial harvesting), decomposers, and so on

A low energy flow results when solar energy that could support life anywhere along the line doesn't.

Energy Flow Demonstrated

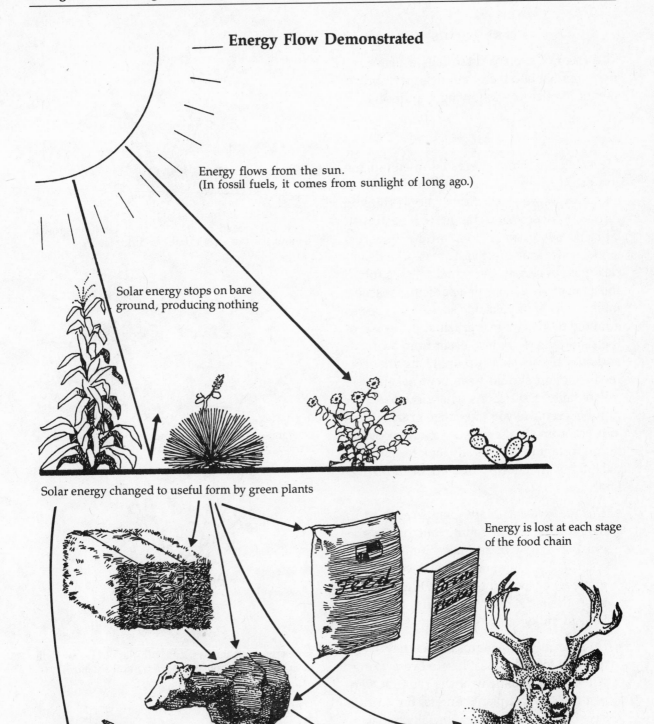

Energy flows from the sun.
(In fossil fuels, it comes from sunlight of long ago.)

Solar energy stops on bare ground, producing nothing

Solar energy changed to useful form by green plants

Energy is lost at each stage of the food chain

A mathematical value for energy flow would be the sum of the calories preserved in organic matter at each level of the food chain from an original amount delivered to a unit of ground over a unit of time.

Plant Forms

The form of a plant often tells a history of management and therefore suggests improvements. Notice the following conditions.

Overgrazing

Livestock operators interested in enhancing the productivity of grassland must learn to spot overgrazed *plants.* (Most people think in terms of overgrazed ranges or pastures.) Chronic overgrazing does produce successional shifts that affect vast areas—the disappearance of certain perennial grasses (often those most productive in a particular season), infestations of unpalatable species, or species adapted to survive overgrazing. But some of the same symptoms also occur from overrest, and most do not show up until long after corrective action should have been taken.

Especially in situations where low paddock numbers require you to change grazing periods according to the speed of plant recovery, it's important to continually look for individual plants that have suffered overgrazing. Look for:

- *Distorted growth:* Many plants respond to overgrazing by flattening out below the grazing height of animals. Some species such as blue gramma may develop into matted, sod-bound patches. Other species raise their defenses by hiding new leaf behind spiky stalks or thorny leaves.

- *Dead Centers:* Bunchgrasses, as they sacrifice root to replace leaf lost to overgrazing, will die back, usually at the center. Overrest too can kill plant centers, but in this case the old oxidized leaf usually remains.

Overbrowsing

Overbrowsing is closely akin to overgrazing, but in this case it affects forbs, shrubs, and trees. Look for:

Some overgrazed plants hide their tender regrowth behind balls of sharp stalks and dry leaves.

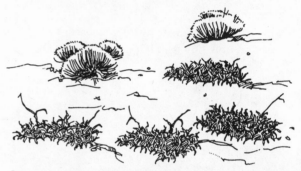

Overgrazed blue gramma and many runner grasses form dense mats.

A healthy bunchgrass plant with time to regrow will have dense lush foliage and a natural shape.

Prolonged overgrazing creates dead centers and prostrate growth around the edge of some plants.

Overbrowsed branch   Browse line

Overbrowsing distorts the Christmas tree shape of pine seedlings.

Overrested plants commonly have gray shocks of old growth and dying centers surrounded by struggling new leaves.

- *Distorted growth/hedging:* Repeatedly bitten branches often develop knobs at the end where new sprouts make a dense cluster. (Gardeners exploit this trait to create hedges by repeated clipping.) Species with a low growth habit may have knobby stems much thicker than normal for the amount of visible foliage. Straight-growing plants such as young pines may split and take the form of bushes (thereby ruining forever the possibility of harvesting timber). Leaves may hide behind spines or old twigs or lie flat against the bark.

- *Browse lines:* Trees lose all foliage below the reach of animals and look like they were trimmed for the benefit of strollers in a park. Some plants may show the knobs and bristles of overbrowsing on their lower branches but long plumes of growth above the reach of animals.

Overrest

Overrest, followed by overgrazing and overbrowsing, most frequently causes degeneration of brittle-environment range and pasture land. Again, successional shifts toward woody species often result. In brittle environments, stagnation often leads to widening plant spacings and loss of organic matter. You should spot the problem *before* these changes occur. Look for:

- Old growth that remains standing into the next growing season or longer—becoming gray or even black in severe cases.

- Plants with dead or weakened centers that have obviously not been grazed recently.

- Weakened root systems on plants that have obviously not been grazed (old growth present). Often these plants can be pulled up easily by hand.

Soil Capping

As the textbook explains in great depth, the capping of bare soil by a hard crust, often cemented by moss and algae, can inhibit succession, and block the cycling of minerals into the soil. Though capping may reduce erosion at the site, it increases runoff that causes erosion elsewhere. Capping may form even on sandy soils, but heavier soils that have lost their crumb structure and lack litter and humus may seal the instant water strikes. In brittle environments, herd effect is usually the most practical way to break up capping while incorporating organic material into the soil.

Soils covered by litter and vegetation will not develop capping. Except where the cover consists of sod-bound grasses (see discussion under Capping on page 110), this presages good water and mineral cycles and advancing succession.

In monitoring we note four degrees of capping:

- *Mature capping* results from long rest. Low-successional lichen, moss, and algal communities often give it a dark tone, and it may sound hollow when tapped.

Mature capping

- *Immature capping* has been broken in the past or scoured by erosion, but it's still strong enough to inhibit succession, water penetration, and aeration.

Immature capping

- *Recent capping* is the result of recent precipitation over broken soil surfaces. It is usually quite thin, depending on soil type, and if left undisturbed becomes immature and eventually mature.

- *Broken capping* is the result of recent animal impact or other disturbance that opens soil to moisture and seed germination.

Recent capping   Broken capping

Identifying Species

How much work should you put into learning the names of plants and animals? Like a

lot of other issues in holistic management, the right answer really fits a different question. Some people memorize names like baseball cards. Don't waste your time doing that. Just learn as much as you can about plants and animals and you'll soon identify them easily.

All organisms fill certain niches in the ecosystem. The more you discover about them and how they fit together, the more you'll understand the dynamics of your land. To help you distinguish one plant from another in order to monitor them, you will of course need names—but they're just the doorknob to more useful knowledge.

Identifying plants from a field guide or key without long practice is a slow and tedious business and misses the point. Get some ex-

perts in birds, insects, mammals, and plants to tour your land with you. Keep notes. Ask questions about what various species need in order to thrive, how they reproduce, what preys on them, and so forth. Though it might take a lifetime to learn a whole biological discipline, you can make a good start on your own limited habitat in a very short time. From then on you can teach yourself what you need to know by your own observations and occasional correspondence.

I once knew an elderly Navajo herder whose practical understanding of his surroundings was breathtaking. He referred to other living things as "plant people" and "animal people" and often talked to them conversationally. People of my background are usually too self-conscious to practice such humility openly. Nevertheless, through that degree of personal intimacy, the grand ideas of holism begin to make sense.

Low stock density is the most frequent problem seen in moderate- to high-rainfall areas.

Grazing Patterns

Overstocking vs. Low Density

Extremely severe grazing was covered earlier as a sign of possible overstocking. Aside from the more complex matter of assuring dormant-season reserves, you have immediate problems if your animals gnaw grass right down to the quick and pick up litter during the growing season. If this happens in paddocks that also contain untouched, rank forage, the problem is probably low stock density.

If livestock are spread too thinly to affect many plants or areas in one grazing, untouched areas may be inedible by the time they return—thus increasing pressure on the remaining space. After several grazing cycles, a large proportion of a paddock may drop entirely out of production. In a nonbrittle area, the ungrazed patches may advance well along the successional path back to forest.

The signs are easy to spot (and are sometimes called "patch" or "all-or-nothing" grazing):

Low-density grazing

- Sharply defined ungrazed patches—large or small, of separate species or the same species

- Extreme grazing of other areas that in non-brittle environments often acquire the clipped look of putting greens

In brittle and low-rainfall areas, low-density grazing patterns are often inevitable and even desirable. Since the overgrowth usually cures well and is an important source of off-season forage, the patches disappear before the next year's growth.

In brittle high-rainfall areas and nonbrittle environments, the overgrowth is typically fibrous, low in nutrients, and must be mowed, burned, or trampled to maintain the pasture. The extreme stock density necessary to control this problem in some areas astounds many people new to using animal impact as a tool to meet their landscape and production goals.

Habits and Routines

Livestock learn routines rapidly—and have a knack for training stockmen to follow them. You must monitor yourself and your land for signs of destructive routines.

Sooner or later cows discover that by bawling loud enough at the gate to the next paddock, someone will assume they are starving and let them through. In times of drought, especially, this can lead to disaster, because you may have to stress animals a wee bit to keep the recovery periods long enough. Giving in to them and knocking a day or two off each grazing can greatly reduce forage and play havoc with your plans and wallet. Faster moves mean overgrazing which means less production which raises pressure for faster moves and yet worse overgrazing in a downward spiral.

More subtle routines can also cause problems. If livestock know they will always move into an adjacent paddock, they soon get the habit of crowding that fence line in anticipa-

The Ivey Matthews Solution

Ivey Matthews, a South Carolina beef producer, puts 464 SAU through 45 paddocks averaging 18.5 acres for a density of about 25 SAU per acre. During rapid growth, however, whole paddocks will become too rank before he gets through all 45.

To avoid this trap, Matthews takes several paddocks out of the cycle and cuts them for winter hay (a variation on the low-rainfall, brittle-environment scenario where the grass will cure well uncut). This strategy allows Matthews to maintain high densities while keeping recovery periods suitably short in the remaining paddocks.

Withdrawing 11 paddocks for hay allows Matthews to keep high density and one-day grazing periods for 33-day recoveries.

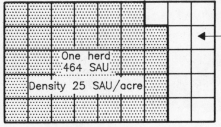

Grazing all paddocks one day gives too much recovery—45 days.

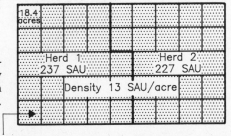

Two herds would mean 2-day grazing periods at half the density for 30-day recoveries.

tion. Stock rotated mechanically around a radial cell frequently produce a pattern of severe grazing along one side of each paddock.

Trailing represents another result of habit. Sometimes the placement of fences causes trails to develop, but trails also result from routines that you can break by moving stock through different gates, changing the location of supplements, or moving a section of fence.

The point is this: Look for signs of destructive routines, and vary them in your biological planning.

Living Organisms

Monitor living things, animal or vegetable, that help or hurt you and try to understand them in context of succession. This perspective enables you to use living organisms as a tool—which is nothing more than influencing succession to your advantage. This is perhaps the most important and least understood concept in holistic management. Changing the environment by direct action (planting, exterminating, stocking, and the like) more often than not fails for trying to produce effects without considering causes.

The textbook explains how such headaches as noxious plant invasions, insect outbreaks, and poor soil aeration will not yield unless you correct the distortion of succession that produced them. The possibilities are limitless. Informally monitoring small aspects of succession can often lead to big advances in management.

A Routine Grazing Pattern

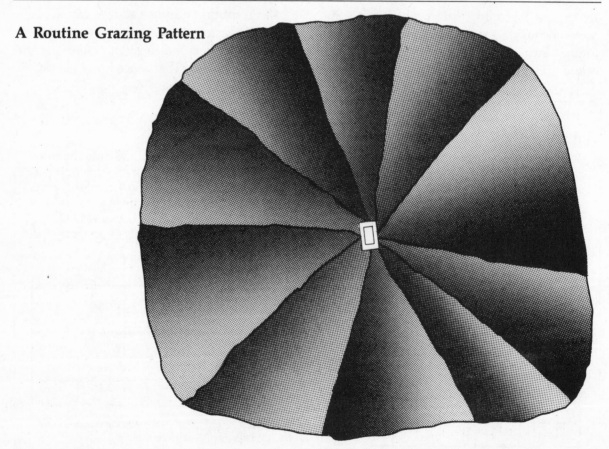

Counterclockwise moves have trained the herd to crowd the fence in anticipation of moves.

A Deeper Look at Living Organisms

There are many examples of creative innovation in land management—from turning pipe leaks into habitat for grasshopper-eating toads to leaving strips of uncut hay for quail cover. For a good expression of the spirit of the idea, here is a passage from Oregon rancher Dayton Hyde's marvelous book Don Coyote. *After closely observing a series of events on his land, Hyde rejected a common policy of draining marshes and poisoning predators and rodents. Instead he restored wetland and fed coyotes. In his words:*

"The ancient marshes were coming back: coarse vegetation sprang up almost overnight, creating a whole new habitat for wildlife—no matter that those grasses did not rate well in the county agent's handbook. Cows got fat on them, and that was what paid the bills.

"All day long, the tea-colored marsh waters absorbed the rays of the mountain sun, and all night long they released heat, warming the climate. Where once even sedges and rushes had turned brown with the night frosts, now they grew green and lush, and frogs croaked all night in the warm, moist darkness. For once we had more feed than cattle to eat it.

"In those same marshes, a host of birds—yellowthroats; redwinged, yellow-headed, and Brewer's blackbirds—raise their young. We helped them by restoring the marshes; they helped us by taking care of insect problems. The great grasshopper epidemics which had plagued my uncle and me became history. In the traditional grasshopper nesting areas, clouds of birds worked on nymphs, carting them away to feed their young. Hour after hour, day after day, sandhill cranes stood shredding young grasshoppers with their long bills.

"We had once relied on poisons; now as we checked the areas where grasshoppers had once turned the land to dust, we found bluebirds, wrens, flickers, sparrow and red-tailed hawks, coyotes, badgers, foxes, cranes, blackbirds, meadowlarks, and even ducks cleaning up the areas. By the time their young were raised, the 'hoppers they'd missed wouldn't have filled a tobacco can.

"There is more to maintaining wetlands than backing up water and letting it stand. Marshes tend to choke up with coarse vegetation, which, left ungrazed or unharvested, rots. Periodic drying is essential.

"The balance of wildlife species too is important. When we lost our coyotes to poison, the raccoons proliferated until there was hardly a nest on the place they hadn't destroyed. Before the coyotes remultiplied and were again numerous enough to control the raccoons, waterfowl production had been almost nil. Now we had ducks and geese back nesting by the hundreds.

"I thought of other species on the ranch. Without flickers, badgers, trout, deer, or chipmunks, the ranch still would have flourished. But if I took away the coyotes, the whole system fell apart. They were as necessary to the well-being of Yamsi Ranch as any tool I owned, including shovel, pick-up truck, mower, hay baler, fencing stretcher, pliers, welding outfit, saddle horse, saddle, rope, medicine, and tractor. In fact, if I were to design a kit for the beginning rancher, a pair of coyotes would have to be included.

"There was a difference between my ranch and every other I knew about whose owners complained about coyote damage. Acre for acre, I had three to five coyotes to their one. Yet while they lay awake at night waiting for predators to kill their livestock, I slept like a baby, not hoping, knowing my calves were safe. The secret, of course, was that I kept my coyotes fed all year round.

"Ever since the great mouse epidemic, I had relied less and less on poisoned grains to keep ground squirrels in check, and now I used none at all. The epidemic of these little grass eaters I had feared didn't happen. Populations stayed low and fairly stable, and I came to look on the ground squirrels I had on my land as beneficial, since they fed my varied predators."

CREATING YOUR PLAN

The following procedure for periodic monitoring of selected sites—the transect procedure—was developed over a number of years by Kirk Gadzia at the Center for Holistic Resource Management. It's presented here in his words along with many warnings resulting from his experience.

Although it contains elements from other monitoring techniques, the transect procedure puts special emphasis on soil surface conditions and plant density. It is here that you will get the earliest warning of a shift in succession—which is far more important than knowing if a shift has already occurred. Monitoring should always look as far down the road as possible for any curves that might require action on your part as a manager.

If you haven't ever done it before, taking the responsibility for monitoring your land yourself should symbolize a significant shift in your whole approach to management. Almost all stock growers weigh their animals at least once a year. When you realize that the stock functions only as a broker in the marketing of solar energy, it makes even more sense to "weigh" the primary agent in this transaction—your land. And just as you would never consider calling in a stranger to weigh your stock, you shouldn't trust the monitoring of your land to anyone but yourself.

The procedure described here takes a few days of hard work once a year and is an essential part of the monitoring process. The hard data yielded by systematic use of this technique do not serve as a substitute for the day-to-day monitoring of growth rates or the general awareness of events that were described in the previous section. In some ways this information is more important because it generates a cumulative written record expressed in numbers that quickly reveal trends.

Gathering the Data

All About Transects

The word "transect" means "crosscut"—the theory being that samples taken from selected crosscuts of land, voters, or any other entity too big to test in its entirety will yield a good approximation of the truth. When you sample random points along a transect of land, it's like polling a random selection of people along a selected street with a prepared questionnaire.

Instead of direct statistical analysis, though, the data collected through this procedure are interpreted through the HRM model. This process will increase your powers of observation and ability to think in holistic terms. However, the conclusions reached by this approach do not always parallel the results of the most frequently practiced range monitoring procedures.

Point sampling has three requirements: (1) a statistically adequate number of points; (2) randomness in choosing points; and (3) points that really are points. (Measuring from a broad mark such as a footprint, you could fudge.)

Obtaining an adequate sample depends somewhat on range type. The procedure outlined here should allow you to measure enough transects within a reasonable time. Experience and statistical evaluation will help you determine the right sample size for specific range types.

To ensure that the sample is random, you will choose the sample points along transects by throwing a dart backward over your shoulder—but even then you must be careful not to aim it. The dart technique gives you a "dimensionless" point, but it's difficult to apply where plant spacing is tight, a common situation in nonbrittle environments. Purely annual ranges also present some difficulties.

Normally you will sample 100 points along the transect at each location. This usually gives reliable results besides making it possible to read many of the findings directly from the data as percentages.

The Monitoring Data Form, shown on page 114, has four major divisions. In the first, you keep a cumulative tally of certain data from all the point samples. The three divisions to the right provide space for specific data about the perennial plant nearest each sample point. Each division has lines numbered from 1 to 33, giving 99 lines total. The 100th point goes on an extra line at the bottom. The rest of the space is for totals and analysis.

Equipment

You will need the following items for your monitoring expedition:

1. Monitoring forms and instruction sheets

2. Clipboard

3. No. 2 pencils or mechanical pencil

4. Measuring tape or ruler

5. Fishing weight on a length of cord (if in brushy country)

6. Pocket calculator

7. 35mm camera and color film (slides or print)

8. Steel posts or suitable permanent markers

9. Bright-colored darts—fitted with long tips

10. This book

Location of Starting Points

Locate permanent starting points for each transect. The transect itself will consist of 100 random points near this point, but the starting point must remain the same from year to year.

For a fairly uniform piece of land you may want to select the permanent sites themselves in a random way. Or you may wish to select areas of particular concern or areas that represent typical conditions. The more uniform the land, the fewer the transects necessary. On many ranches a minimum of three to five transects per grazing cell give good information. You must balance the time you can invest against the precision of information you need.

Once you've located the starting points on the land, mark them with a steel post, rock pile, or other permanent fixture. Record the locations on the map and write down the directions for getting there if at all complicated. Mark the routes on the map if necessary.

Permanent transect markers on a brittle range in New Mexico

Time of Year

Experience has shown that monitoring yields the best information during the most active growing season. This way you have the best chance of detecting overgrazing and distorted growth form. It's also the best time to identify many plants and observe their characteristics. Biological activity is normally highest as well.

The main point is that monitoring should be done at the same biological time each year. The exact day is not critical since seasons shift from year to year. Don't schedule it when you know you'll be cutting hay or wheat. Maybe right after. And try to have the same person run the transects each year. If you must use someone new, work with them ahead of time to standardize the procedure. New personnel will inevitably change procedures slightly, but you can minimize the problem.

Photographic Record Keeping

Take at least two photos at each permanent transect marker.

Standing at the permanent marker, aim the first photo in the direction of the transect. Although the location instructions for each marker should indicate this as a compass bearing, the photo should include a recognizable feature—say, a feature on the horizon—to show it is on the mark. A second permanent post, marked in intervals to show scale and placed about 20 feet away, works also.

The photo should give a good view of ground cover and include as little sky as possible. Try to shoot it at the same time of day each year so the light is the same. If your camera adjusts, set the lens opening at f16 or smaller for maximum depth of field and focus on a middle distance. (If that makes the shutter speed slower than 1/125 sec, you'll have to compromise or use a tripod.)

The second photo is taken from 4 feet aboveground and aimed straight down to give a good view of the soil surface. Because a marker post may attract animals, take the shot a few yards away along the transect. A short, permanent steel peg that will show up in the lower right corner of the photo will ensure that you get the same plot every time.

Take as many pictures as necessary to ensure one good picture. A databack attachment for the camera will permanently record the date directly on the slide or photo. Otherwise write the date of the transect on a card and put it in the picture. In any case, process the film as soon as possible and immediately record the information on the photo itself. Store the labeled prints or slides in plastic looseleaf binder pages (available at any photo shop).

Aside from these two obligatory shots of each transect, you'll also want to document details at individual sample points, such as examples of mature capping, new seedlings, or unknown plants. A note of the subject and the picture number on the form will help you sort the photos later and keep them organized.

Photographic records are an invaluable tool to help you *see* changes on the land over time. Be thorough and keep them clean and neatly organized.

Transect Information

Fill in all the required information about each transect at the top of the Monitoring Data Form. "Date HRM" refers to the date when holistic resource management was begun. Use the note column or back of the sheet to record any other information needed to identify that particular transect.

If forms are shuffled out of order or evaluated by someone else, lack of proper identification will render them meaningless. Complete information will also ensure easy retrieval from whatever filing system you use. Because of the record's value, it's a good idea to file a copy somewhere safe from fire.

Point Hits

After selecting the starting point and taking photographs, you can begin monitoring fieldwork. Stand at the permanent starting point and toss the dart over your shoulder in the direction of the transect. Painting the dart fluorescent orange or tying on a piece of orange flagging may help you to find it in dense cover. Short tosses are easiest to find and don't bias the results.

Making a Sampling Dart

A dimestore dart makes a classy tester that pierces tall grass. Yank out the point. Drill a 5/64-inch hole. Screw in a bicycle spoke—cut to 8 inches and sharpened.

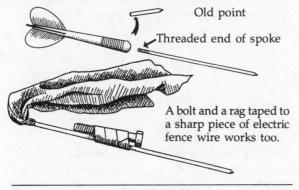

Old point

Threaded end of spoke

A bolt and a rag taped to a sharp piece of electric fence wire works too.

Accuracy depends on tossing randomly. Make a conscious effort not to throw the dart where *you* hope it lands. The over-the-shoulder toss helps. If the dart doesn't stick, sight straight down over the point to the spot below. When you finish documenting one sample point, stand there, back to the transect direction, and toss again. Depending on how far you toss, the whole transect will run 200 to 300 yards.

Keep track of the numbers at each point by the dot tally method.

$\vdots$ = 4   = 6   = 9   = 10

Ground Cover

Cover is the first category in the left-hand column of the Monitoring Data Form. Use the dot tally to record what the dart hit. "Bare" (for bare soil) and "rock" speak for themselves. "Litter 1" is a new, undecayed layer of litter (leaves, sticks, dung). "Litter 2" indicates deeper litter that is decaying and being incorporated into the soil. Often the litter grades into the soil without a distinct boundary. "Basal" refers to basal cover—the area actually covered by the root crowns or stems of perennial plants. If the dart strikes basal area, add to the dot tally for "basal" and record the distance to the nearest plant as zero in the right-hand columns for that point sample.

In most bunchgrass-dominated communities, basal hits rarely exceed 10 to 15%. In areas with sod-forming grasses, individual basal areas are hard to distinguish and you must devise a standard. One way is to record a basal hit only if the point falls into a living clump. Otherwise record litter or bare ground and measure the distance to the nearest live shoot of the sod-forming grass. If the dart lodges in vegetation aboveground, use the fishing sinker and cord as a plumb bob to find a point on the soil surface directly below and treat that as a hit.

Locating a Point with a Plumb Bob When the Dart Gets Caught

Canopy Type

Canopy refers to any part of a plant an ant would see looking straight up from the ground. It is everything that would cast a shadow if the sun were straight overhead. Canopy helps shield the soil from raindrop impact and shades it from direct sun and wind.

Record canopy only if it's directly above the point hit. You may have to use the string and plumb bob to check this. You may have a basal hit and no canopy or a basal hit and canopy from the same plant.

Grazing greatly influences the aerial cover of grasses and forbs and to some extent brush also. If last year you monitored just before grazing and this year just after, you'll note a great change. You might want to discount this information in your analysis, but it shouldn't affect your conclusions, as soil cover and plant density are far more stable indicators of community health.

Capping

Soil capping has a major influence on progress toward your landscape goals. Breaking up capping is one of the primary steps toward better water and mineral cycles, energy flow, and higher succession. Sample the capping within a 6-inch radius from the dart strike. If that doesn't work for you, set a different standard—but keep it constant.

The previous section (page 100) describes the four levels of capping listed on the Monitoring Data Form. Since defining these levels is somewhat subjective, photos and training are essential. Particularly important is the handling of the sod-bound conditions often associated in North America with blue gramma in many brittle environments and Bermuda grass (*Cynodon dactylon*) in less brittle ones. The latter is known as couch grass in Africa and fills the same niche. Thus what

you might interpret as "covered soil" is in fact not so good—a strong argument for adding a separate category. If sod-bound strikes would make up a significant part of the sample, record them as "covered" but keep a separate dot tally in the notes section to help you make a better analysis later.

Living Organisms

Because succession includes all biological forms, not merely plants, this category gives some indication of the general life in your soil. Although the measure is imperfect—you can't assess underground activity accurately—it counts for a lot.

Check within a 6-inch radius around the strike point (or other standard size if you choose), and note any signs of large or small animal activity—tracks, droppings, burrows, mounds, worm castings, or actual sightings. The Monitoring Data Form has four categories, and the dot tally in each will help you determine the level of succession.

Plant Type

Locate the nearest *perennial* plant and decide if it's a grass, forb, bush, or tree. Forbs are any flowering plants, other than grasses, that do not develop woody stems. Brush may be simply scrubby tree growth. Fix your own definition according to the expected growth form of the species in your area. A photo may help here. Keep the dot tally for all four groups by the "Plant Type" heading.

Habitat

Habitat here refers to the soil category to which your sample plant is best adapted. Dry (xerophytic), middle (mesophytic), and wet (hydrophytic) plants are described on page 95.

Dot Tally for Dart Throws

Ranch: RIVER BEND		Cell: 2	
Nrst. Plant & Point Hits		Dot Tally	Total
Cover	bare	⊠⊠⊠⊠⊠⊠⊡ ∶	56
	litter 1	⊠⊠⊠	30
	litter 2	⫶ ∶	5
	rock	•	1
	basal	⊡	8
Canopy	grass	⊠⊠⊠⊡ •	14
	forb	• •	2
	brush	⊠ YUCCA	9
	tree		
Capping	mature	⊠⊠⊡ ∶	24
	immature	⊠⊠⊡ ∶	24
	recent	⊠⊡⫶	17
	broken	⫶⫶	7
	"covered"	⊠⊠⊡⫶	28
Living Organ-isms	insect	⊠⊠⊡	22
	bird	• • •	3
	small animal	⊠	9
	large animal	⊠⊠⊠⊠⊠⊠⊡⫶	61
Plant Type	grass	⊠⊠⊠⊠⊠⊠⊠⊠⊠	90
	forb	• • ∶	4
	brush	⫶ ∶	6
	tree		
Habitat	dry		3
	middle	⊠⊠⊠⊠⊠⊠⊠⊠⊠⊠⫶	97
	wet		

Distance

Next measure the distance from the dart strike to the base of the plant and record this under "Dist." on the data line for each point sample with a letter indicating plant type—say F 1.2 for a forb 1.2 inches away, G 12 for perennial grass 12 inches away, and so on.

If an *annual* plant is closer than the perennial, put a simple check by the distance to the perennial (✓ G 12). (Annuals—sunflowers and cheat grass, for example—grow from seed every year.)

In the sod-bound communities mentioned earlier or where a pure stand of dense grass covers the ground, some people record the point strike with a check and a letter for the species and then measure the distance to the nearest perennial of another species, which may be quite far.

Baseline plant-spacing information does not in itself tell you much, but comparisons to later monitoring will provide one of the most sensitive indicators of a trend. Expanding bare areas and falling plant density are signs of a declining ecosystem. Decreasing distances indicate the reverse.

In brittle environments, a negative trend usually means a lack of animal impact. The most dramatic improvements will come from an increase in seedlings, but distances also decrease as basal crowns mature and enlarge. The record of age classes will show which process dominates.

Cool- and Warm-season (C/W) Plants

A mix of cool- and warm-season plants, as we have seen, is evidence of higher succession. Because these proportions are extremely sensitive to overgrazing, changes can have a big impact on management and planning.

If most of the moisture falls as snow or rain during the cooler months of the year, cool-season plants should predominate. Conversely, wet summers and dry winters and springs normally give warm-season plants an edge. Some areas have split seasons in which both types grow well and nearly balance. Grazing, however, can radically distort all these proportions. Even when cured forage of one type is adequate, livestock will select new growth of the other and may overgraze it to extinction if permitted. If either grazing or climate limits one type of plant, exceptional weather may induce dramatic blooms of annuals that fill the niche.

The easiest way to distinguish between these types of plants is to observe the time of year when they actively grow. Plant experts usually know the growth habits of the species.

Age

The proportion of plants in each age class strongly indicates the progress of succession. Young plants and seedlings determine the plant community of the future. But since they are relatively inconspicuous at exactly the time they should influence your management decisions, ground-level monitoring is crucial. If, for example, you find a large number of brush seedlings that don't fit your goal, the sooner you can respond the better.

As noted before, closer plant spacing can result either from new seedlings or the growth of basal cover by maturing plants. By noting the age class on the Monitoring Data Form, indicate which is the case:

S = Seedling

Y = Young

M = Mature

D = Decadent or dying

R = Resprout

Form

Form refers to the plant's shape under the influence of grazing or rest. (See pages 96-99.) Use the following notation:

N = Normal and vigorous. This is usually obvious, but look for seed production, tillering (new stems at ground level) and branching, and lack of old stale growth.

Or = Overrested.

Og or Ob = Overgrazed or overbrowsed.

D = Dying. No evidence of overgrazing or overresting, but the plant is dying—maybe because of old age, excessive erosion, insect damage, trampling, or water-table changes.

Species

If you can identify plants by name, do it here. This ability is not an end in itself, but combined with knowledge about where various species fit into the ecosystem it will add a lot of substance to your monitoring.

Because space is limited on the Monitoring Data Form, use an abbreviation for the species name. The abbreviation can be any designation you like, but list the full names with the abbreviations alongside so you can identify them in the future. (The U.S. Soil Conservation Service and other government agencies have standardized abbreviations for plants.)

The back of the form has space for analysis by name and type and for figuring the proportions of each. Write the names of the species under the proper heading (grass, forb, and so on), count the number of times each occurred, and note the percentage in the right-hand column. If you have 100 samples, the number will be the percentage.

Totaling the number of species will give you a good idea of the complexity of the plant community. Totaling the number of plants within each species gives you an idea of the relative composition of the community. As succession advances, you'll probably note increases in the number of species and decreases in the number of individuals within dominant species.

If you can identify other plants besides those that turn up in the point sampling, note them on the back of the form. This will give

Data on Perennial Plant Nearest Sample Point

Xsect. nos. 1   Photo nos. 6/20/88 - 9 - 15   Examiner:

Dist.	C/W	Age	Form	Specie	Dist.	C/W	Age	Form	Specie
0.1	W	M	N	BL.GR.					
√3.0	W	M	N	"					
1.0	W	M	N	"					
.25	W	M	N	"					
0.5	W	M	N	"					
1.5	W	M	N	SIDE					
0.7	W	M	N	BL.GR.					
√2.5	W	M	N	SIDE					
1.7	W	M	N	"					
.75	W	Y	N	"					
2.25	W	Y	N	"					
1.4	W	Y	N	"					
0.7	W	M	N	BL.GR.					
0	W	M	N	"					
1.0	W	M	N	"					
0.7	W	M	N	"					
0	W	M	N	"					
√2.5	W	Y	N	BLK.GR.					
0.7	W	M	N	BL.GR.					
1.0	W	M	N	"					
0.2	W	M	N	"					
0.5	W	M	N	"					
0.7	W	M	N	"					
1.0	W	M	N	FL.WIT.					
.75	W	Y	N	S DROP					
1.5	W	Y	N	3 AWN					
0.5	W	M	N	BL.GR.					
0.7	W	M	N	"					
3.0	W	M	N	"					
1.0	W	M	N	"					
1.25	W	M	N	SNAKE					
√5.5	W	Y	N	BL.GR.					
√2.0	W	Y	N	SIDE					
1.0	W	M	N	BL.GR.					
41.85	W 34	(Y 8) (M 26)	(N 34)						

you a better indication of diversity but will not figure in the percentages.

Erosion Class

Erosion in general was discussed on page 92. At this level of monitoring, rate it by observation as you move along the transect according to the table shown here. Again, your judgments will be somewhat subjective because the categories overlap. Flow patterns may be downstream evidence of soil movement or sheet erosion or rills.

Rate the transect area at least twice. After the first 33 points and again after the second 33 are good times to take a break and make notes anyway. The two scores can be averaged in the summary. There is also space for a third rating after the last 34 points if erosion characteristics vary significantly.

Totals and Percentages

Totals from the dot-tally portion of the Monitoring Data Form go to the right of the dot-count column. Data on the 100 nearest plants are subtotaled at the bottom of the form.

The section for erosion has blanks for totals for each erosion class and each of the three points on the transect (see page 114). These

Erosion Condition Guide

CLASS	1	2	3	4
Soil Movement	Little evidence of movement of soil particles.	Moderate movement is visible and recent. Slight terracing evident.	Soil movement with each event deposited against obstacles.	Subsoil exposed over much of the area.
Surface Litter	Litter accumulating and incorporated into soil. Rapid breakdown.	Litter movement by wind and water evident. Slow breakdown.	Extreme movement, large deposits against obstacles.	Litter removed by animals, wind, and water.
Pedestals	No evidence of pedestals.	Small plant and rock pedestals in flow patterns.	Most plants and rocks on pedestals. Some root exposure.	Plant roots exposed. Most areas affected.
Flow Patterns	Little evidence of particle movement.	Well-defined; small and intermittent.	Flow patterns evident with deposition of soil, litter, and fans.	Patterns numerous and noticeable. Large barren fan deposits.
Rills	Small rills absent or at infrequent intervals. Vegetation present in depression.	Rills from 1 to 8 inches wide occur in exposed areas, at frequent intervals, greater than 10 feet apart.	Rills occur frequently and quickly cutting sides; often exposing roots.	Rills joining at short intervals and denuding large areas.
Gullies	May be present in stable condition. Slopes and channel generally stable and vegetated.	Gullies present but with slumping sides from animal impact, vegetation becoming established.	Gullies numerous and well-developed. Active erosion evident along 25 to 50% of their length.	Sharply inclined gullies with little vegetation cover large areas and are actively eroding over 50% of their length.

Completed Monitoring Data Form

Center for Holistic Resource Management — MONITORING DATA FORM

Ranch: RIVER BEND   Cell: 2   Xsect. nos. 1   Photo nos. 6/20/88-9-15   Examiner: K. WILLIAMS   Date: 6/20/88   HRM 2 YRS. 6 MO.

Nrst. nos. 1

Left summary — Nearest Plant & Point Hits

Category	Subtype	Total
Cover	bare	56
Cover	litter 1	30
Cover	litter 2	5
Cover	rock	1
Cover	basal	8
Canopy	grass	14
Canopy	forb	2
Canopy	brush	9
Canopy	tree	
Capping	mature	24
Capping	immature	24
Capping	recent	17
Capping	broken	7
	"covered"	28
	insect	22
Living Organisms	bird	3
Living Organisms	small animal	9
Living Organisms	large animal	61
Plant type	grass	90
Plant type	forb	4
Plant type	brush	6
Plant type	tree	
Habitat	dry	3
Habitat	middle	197
Habitat	wet	
Erosion	soil	6
Erosion	litter	6
Erosion	pedestal	5
Erosion	flow	5
Erosion	rill	3
Erosion	gully	2

Erosion summary (C/W | Age | Form): soil 3 | 2 | 1; litter 2 | 2 | 2; pedestal 2 | 2 | 1; flow 2 | 2 | 1; rill 1 | 1 | 0; gully 2 | 0 | 0; Total 13 | 9 | 6

Overall tally: C 22 Y 26 OR 22 / W 92 M 69 N 46 / S 5 S 5 — Grand Total / Subtotal — 1:12

Transect measurement data (Distance / C/W / Age / Form / Specie)

Pt	D1	C/W	Age	Form	Specie	D2	C/W	Age	Form	Specie	D3	C/W	Age	Form	Specie
1	.1	W	M	N	BL.GR	0	W	M	N	BL.GR	1.0	W	Y	N	BL.GR
2	√3.0	W	M	N	"	0.2	W	M	N	BL.GR	0.4	W	M	N	"
3	1.0	W	M	N	"	0.3	W	M	N	SNAKE	3.2	W	M	N	BLK.GR
4	.25	W	M	N	"	1.0	W	Y	N	SIDE	0.5	W	M	N	BL.GR
5	0.5	W	M	N		2.3	W	S	N	BL.GR	1.0	W	M	N	TRIDENS
6	1.5	W	M	N	SIDE	0.4	W	M	N	HAIRGR.	√2.8	W	M	N	BL.GR
7	0.7	W	M	N	BL.GR	2.5	W	S	N	SNAKE	1.1	C	M	N	BUFFGR
8	√2.5	W	M	N	SIDE	2.0	W	S	N	BL.GR	√2.0	C	M	N	SOTAIL
9	1.7	W	M	N		0.9	W	M	N	BL.GR	0.8	W	Y	N	BL.GR
10	.75	W	Y	N	"	0.9	W	M	N	"	2.0	W	M	N	YUCCA
11	2.25	W	Y	N	BL.GR	1.0	W	M	N	HAIRGR.	1.9	W	S	N	BL.GR
12	1.4	W	Y	N		0.2	W	Y	N	"	0.4	W	M	N	"
13	0.7	W	M	N	BL.GR	0.2	W	S	N	"	1.2	W	S	N	"
14	0	W	M	N	"	1.8	W	Y	N	"	0.2	W	Y	N	SIDE
15	1.0	W	M	N	"	2.5	W	Y	N	"	0.2	C	Y	N	FEATHER
16	0.7	W	M	N		1.0	W	Y	N	PERFORB	1.2	W	M	N	HAIRGR
17	0	W	M	N		1.0	W	Y	N	PERFORB	2.0	W	M	N	3-AWN
18	√2.5	W	Y	N	BL.GR	0	W	M	N	BL.GR	1.5	W	M	N	SIDE
19	0.7	W	M	N	BL.GR	0.7	W	Y	N	SIDE	0.2	W	M	N	HAIRGR.
20	1.0	W	M	N	"	1.0	W	M	N	BL.GR	0	W	M	N	SIDE
21	0.2	W	M	N	"	0.2	W	M	N	HAIRGR	0.6	W	Y	N	S.DROP
22	0.5	W	M	N	"	2.7	W	M	N	BL.GR	0.8	W	S	N	RING M.
23	0.7	W	M	N		1.5	W	M	N	SNAKE	0	W	M	OR	RING M.
24	.75	W	M	N	FL.WIT	5.8	W	M	N	BL.GR	1.8	W	M	N	BL.GR
25	1.5	W	Y	N	SDROP	0.1	W	M	N	SIDE	0.6	W	M	N	BUFFGR
26	0.5	W	Y	N	3-AWN	1.5	W	Y	N	BL.GR	0.2	W	M	N	FL.WITCH
27	0.7	W	M	N	BL.GR	1.0	W	Y	N	PERFORB	0.7	W	M	OR	RING M.
28	3.0	W	M	N	"	√1.5	W	M	N	BL.GR	0.4	W	M	N	BL.GR
29	1.0	W	M	N	"	0	W	M	N	"	0	W	M	N	BLK.GR
30	1.25	W	M	N	SNAKE	1.0	W	M	N	RING M.	0.7	W	M	N	BL.GR
31	√5.5	W	Y	N	SIDE	0.9	W	Y	N	S.DROP	0.2	W	Y	N	BL.GR
32	√2.0	W	Y	N	PER.FB.	0.3	W	Y	N	BL.GR	0.2	W	M	N	BL.GR
33	1.0	W	M	N	BL.GR	√1.1	W	M	N	BL.GR	√1.5	W	M	N	BUFFGR
100	41.85					38.6					31.3				

Group bottom tallies:
- Group 1: (W) 34; (Y) 8 / (M) 26; (N) 34
- Group 2: (W) 33; (Y) 12 / (M) 18 / (S) 5 / 3; (N) 33
- Group 3: (C) 2 / (W) 31; (Y) 6 / (M) 25 / (S) 5 / 2; (OR) 2 / (N) 31

Notes

GRASSES
- BLUE GRAMMA (BOUTELOUA GRACILIS) — 49
- SIDE OATS GRAMMA (BOUTELOUA CURTIPENDULA) — 14
- BLACK GRAMMA (BOUTELOUA ERIPODA) — 3
- HAIRY GRAMMA (BOUTELOUA HIRSUTA) — 9
- 3-AWN — 2
- BUFFALO GRASS (BUCHLOE DACLYLOIDES) — 3
- RING MUHLY (MUHLENGERGIA TORREY) — 3
- BOTTLE BRUSH SQUIRREL TAIL (SITANION HYSTRIX) — 1
- FALL WITCHGRASS — 2
- NM FEATHERGRASS — 1
- SAND DROPSEED (SPOROBOLUS CRYPTANDRUS) — 3

FORBS

BRUSH
- SNAKEWEED (GUTIERREZIA) — 4
- TRIDENS — 1
- YUCCA — 1

sums in fact have far less meaning than the individual grades. One serious gully is a major problem, for example, even if the other two evaluation points show none. On the other hand, an extreme figure among the totals for either lines or columns might justify a second look and some deeper analysis.

Total the distances for each section of the form. (The third section will have 34 distances.)

For the C/W, Age, and Form columns, subtotal the number in each category (say, 28 W, 6 C; 2 Og, 5 Ob; and so forth).

Put grand totals for all these items in the spaces at the lower left corner of the form. Because there are 100 points, these numbers will be percentages. The total distance divided by 100 will be the average distance—your density index.

Summarizing the Data

Unanalyzed monitoring is like unsmelted ore—a monument to wasted toil. You must separate the useful knowledge out of all the information you collected on the Monitoring Data Forms. The process demands little computation but a good deal of thought. The Monitoring Summary Form breaks the process into steps that help synthesize information and bring it to bear on the status of your landscape goals. Not surprisingly, the format follows the HRM model, which after all was devised to connect observation and action. The steps are as follows.

Section 1

Fill in the identification information at the top of the form. If your files ever become

MONITORING SUMMARY FORM

1	Ranch:											
2 **DATA SUMMARY**	% Cover	Canopy	Capping	Animal Sign	Plant Type	Habitat	Erosion	Avg. Distance	AGE	FORM	Species
	bare 56	G— 14	M— 24	I— 22	G— 90	Dry— 3	S— 2	1.12	S— 5	N—98	List separately by plant type groupings
	litter 1 30	F— 2	Y— 24	B— 3	F— 4	Mid—97	L— 2		Y—26	OR—2	
	litter 2 5	B— 9	R— 17	S— 9	B— 6	Wet—	P— 2		M—69	OG—	
	rock 1	T—	B— 7	I— 61	T— ●		F—1 2/3	cool— 2	D—	OB—	GRASS 90
	basal 8		C— 28				R— 1	warm— 98	R	D	FORBS 4
							G— 2/3				BRUSH 6
							9 1/3				
3	Production, Land Description, and Quality of Life Goals:										

3 PROFIT FROM LIVESTOCK AND STABLE WATER COURSES   HIGH SUCCESSIONAL OPEN GRASSLAND   COLLEGE FOR THE KIDS, GOOD HUNTING, AND CLOSE COOPERATION WITHIN FAMILY

4 Ecosystem Blocks: Water Cycle; Mineral Cycle; Succession; Energy Flow

SUCCESSION - BETTER DIVERSITY (% OF BLUE GRAMMA FELL 8% BUT TOTAL GRASS UNCHANGED - BLACK GRAMMA NEW, HAIRY GRAMMA INCREASED) SEEDING % UNCHANGED

WATER CYCLE- SIGNIFICANT DECREASE IN MATURE CAPPING, THOUGH EROSION STILL NOT ARRESTED

MINERAL CYCLE-FEW OVERRESTED PLANTS, DECREASE IN CAPPING, SLIGHT DECREASE IN SPACING, BUT A DISTURBING LOSS OF LITTER.

ENERGY FLOW - INCREASE OF BARE GROUND DUE TO LOSS OF LITTER. NO INCREASE YET IN COOL-SEASON PLANTS. BETTER PLANT DENSITY, AND NO EVIDENCE OF OVERGRAZING.

When the data from the samples have been boiled down and compared to data from previous years, the statements in Section 4 become the basis for decisions.

separated, there should be no doubt about which Monitoring Data Forms contributed to the summary.

Section 2

If you're combining data from several Monitoring Data Forms, enter the *averages* from all the numerical categories here. Add the grand totals (final averages in the case of erosion classes) from each form, and divide by the number of forms. (Totals of individual species, as opposed to plant types, will have to go separately or on the back of the form.)

Section 3

You should already have quality of life, landscape description, and production goals for the area being monitored. Writing them down here will keep them in your mind as you consider how the monitoring data can help you achieve them.

Section 4

Here you rate the status of the four ecosystem blocks according to the data summarized in Section 2. Remember: This is not a mathematical process but a subjective analysis. Generally, however, the data relate to the ecosystem blocks in the following ways.

Succession

* *Cover:* Tells much about succession. Obviously an increase in bare ground and rock represents a decline. Litter, especially Litter 2, is often a precondition for advancing succession. Basal cover is one indication of density.

* *Canopy:* Like cover, canopy mitigates the harshness of the surface environment by providing shade and wind protection and slowing the impact of rain. Thus it too helps advance succession. The kind of plants that provide canopy may tell something of the direction of succession.

* *Capping:* This third aspect of soil surface condition also has profound implications for succession. Mature capping is a sure sign of stagnation or decline. Covered soil, (except for the sod-bound condition described elsewhere) is a sure sign of advance. The other three categories fill out the scale.

* *Living Organisms:* Quantity alone, unless combined with diversity, does not indicate an advance. An "infestation" reflects a lack of balance and stability, but it isn't necessarily bad. Certain species will predominate at given levels of succession or moments of time. Massive hatches of cicadas occur every 17 years in eastern North America, for example, but cause little damage where succession has produced diverse predators.

* *Plant Type:* Depending on your landscape goal, changes in the proportions here will show your progress toward achieving it.

* *Habitat:* Interpret the marks here in terms of your goal. "Middle" plants are most productive, and in many cases advancing succession will change habitat to increase them. A diversity of habitat or more of a particular kind may suit your goals better.

* *Erosion:* Erosion and bare ground are key indicators of poor water cycles, but low succession and exposed, eroding soil go hand in hand.

* *Distance:* This index of plant density (except in sod-bound conditions) normally reflects succession. Closely spaced plants hold litter in place and keep soils covered.

- *Cool/Warm:* As an aspect of diversity, changes here are significant.

- *Age:* Obviously the mix and number of plants in the youngest class determine the community of the future. The proportion of moribund plants shows changes from the past.

- *Form:* Overgrazing, overbrowsing, and overrest are three of the most powerful influences on succession. Plentiful signs of any of them are serious warnings.

- *Species:* Consider the general diversity, the proportion of annual plants, and the presence of high-successional plants.

Water Cycle

- *Cover:* Litter or basal cover enhances the water cycle.

- *Canopy:* Shade, wind protection, and anything that breaks the impact of raindrops benefits water cycles. Note, however, that the large drops falling from tall trees on bare ground may be worse than uninterrupted rain.

- *Capping:* Capping is a primary source of poor water cycle.

- *Living Organisms:* Burrowing animals directly affect water cycle. Although more animal activity is usually better than less, evaluation is subjective.

- *Plant Type:* Plants with varying root depths get their water at different levels, so diversity reflects health. Think this through for your mix of plants. What conclusions can you make about depth of water table, transpiration, surface evaporation, and so on?

- *Habitat:* The dry, middle, and wet categories directly reflect water cycle.

- *Erosion:* A direct indicator of water cycle health.

- *Distance:* Changes in plant density may precede changes in water cycle.

- *Cool/Warm:* Not a prime indicator of water cycle status.

- *Age:* Water cycle changes may explain why certain seedlings do or don't survive, why certain classes of plants are dying, and other peculiarities of age distribution. Think it through.

- *Form:* Overgrazing and overbrowsing affect water cycle. Overrest may be a cause of damage. Dying plants may be direct symptoms of change. Think it through.

- *Species:* Certain species associated with good or bad soil aeration may indicate changes in water cycle.

Mineral Cycle

- *Cover:* Litter, especially Litter 2, is a direct measure of improvement. Check how quickly litter seems to be breaking down.

- *Canopy:* Note what happens to leaf fall.

- *Capping:* Capping generally inhibits mineral cycling.

- *Living Organisms:* The decomposers and all creatures that live in the soil or burrow through it are agents of mineral cycling.

- *Plant Type:* Look for plants of varying root depth, and note the general diversity of species.

- *Habitat:* Waterlogged, desiccated, or badly aerated soil inhibits growth and therefore the cycling of minerals. True wetlands, however, typically support enormous organic activity and can cycle a tremendous mass of complex compounds.

- *Erosion:* Any loss of soil is a break in the mineral cycle.

- *Distance:* Greater density implies more cycling and ability to keep soil covered.

- *Cool/Warm:* Good as evidence of organic activity but difficult to evaluate in this context.

- *Age:* Relevant as an indicator of growth activity. Obviously a large proportion of dying plants, if decaying slowly, amounts to a slowing of the cycle.

- *Form:* Overrest is a common bottleneck in the mineral cycle because potential nutrients stay locked up in old vegetation for long periods of time.

- *Species:* Look for diversity and varied root depths.

Energy Flow

- *Cover:* Bare ground or rock obviously indicates the worst possible energy flow.

- *Canopy:* Energy flow is largely a function of the total area of leaf that converts solar energy into forage. What does your canopy report tell of this?

- *Capping:* Capped soil probably indicates less than possible organic growth and soil activity, but it doesn't directly measure energy flow.

- *Living Organisms:* A better energy flow supports more activity, but the relationship is too complex to register reliably in a transect.

- *Plant Type:* Broad-leafed plants obviously indicate more energy flow.

- *Habitat:* As both dry and wet plants usually have narrow leaves, a preponderance of broader-leafed middle plants reflects a good energy flow.

- *Erosion:* As erosion usually means bare ground, erosion does affect energy flow.

- *Distance:* An important indicator. Tighter density means more leaves harvesting sunlight.

- *Cool/Warm:* A good mix of both cool- and warm-season plants means a longer period for converting solar energy and thus a better flow.

- *Age:* Remember that green leaf area exposed to sunlight determines energy flow.

- *Form:* Overbrowsed, overgrazed, overrested, or dying plants obviously don't convert maximum energy.

- *Species:* Species is important as it relates to cool- and warm-season diversity, habitat, and plant type—all covered above.

Section 5

In Section 4 you identified certain problems or changes in the four ecosystem foundation blocks. Here you determine what tools account for these situations. Since the textbook gives a detailed description of their effects and the way they interrelate, it might be wise to review the text before deciding which practices you will alter or which you will continue. If Section 4 identified a poor water cycle because of pervasive capping and absence of litter, for example, you might focus on insufficient animal impact as the prime cause.

Section 6

You have identified the conditions and the causes and thereby the tool that can bring about a remedy. Now you'll have to decide how you will use that tool.

Decide which guidelines apply and how they will govern your actions. When you complete this step, you'll find that you have outlined the action you should take. It remains only to change your plans according-

Analyzing Data and Redirecting Policy

5	Range Influences: Money/Labor (Rest Fire Grazing Animal—Impact Living Organisms Technology) Human Creativity
	LOSS OF LITTER, EVEN THOUGH OTHER INDICATORS FAIRLY GOOD, PROBABLY MEANS STOCKING RATE TOO HIGH THIS SEASON. ANIMAL IMPACT GOOD, LITTLE EVIDENCE OF OVERREST, OR NEED FOR A BURN, OR WEED CONTROL.

6	Guidelines: **Testing:** Whole Ecosystem, Weak Link, Marginal Reaction, Gross Margin, Energy/Wealth, Society and Culture. **Management:** Time, Stock Density, Herd Effect, Population Management, Burning, Flexibility, Biological PMC, Organization/Personal Growth, $ PMC
	REPLAN TO DECREASE SDA HARVESTED EACH GRAZING BY 20% BUT INCREASE DENSITY BY COMBINING FIRST- AND SECOND-CALF HEIFERS AND CUTTING GRAZING TIMES. TO FAVOR COOL-SEASON PLANTS, SCHEDULE SHORT, HIGH-DENSITY GRAZINGS MIDSUMMER AND AGAIN AT THE ONSET OF DORMANCY IN EARLY OCTOBER.

ly. Again the textbook explains in detail how to apply the guidelines.

Returning to the water cycle example in Section 5 where you decided you needed more animal impact, you discover that two guidelines—stock density and herd effect—apply most directly. The textbook will tell you various ways to increase stock density and herd effect, and then you'll have to run each option through the testing guidelines. To improve density, you might choose between increased fencing, amalgamation of herds, or increasing stocking rate. The testing guidelines will help you decide which. To improve herd effect, you could train animals to respond to attractants of various kinds.

A Note on Shortcuts

Make every effort to monitor as thoroughly as possible—especially in the first years. If for any reason you can't complete the whole process in a given year, the following information is most important.

- Fixed-point photos
- Soil surface information
- Distance to nearest perennial plant
- Type of nearest plant
- Erosion class rating
- Thorough notes on observations

SUMMARY

If biological planning is the route to managing your piece of the natural world, monitoring is the best way to learn from it. Monitoring a farm or ranch involves both a constant attitude of openness and curiosity and a self-disciplined labor of measuring, recording, and photographing actual data.

Anyone who has ever contemplated the life of peasants and indigenous people is amazed at their enormous hodgepodge of wisdom and lore and has wondered how they got it. What unfortunate individual discovered—and alerted his survivors to—the toxic powers of the fly amanita mushroom? Why does an ancient Greek manuscript declare that blowfish crawl out of the water and mate with goats at certain phases of the moon? And why does that idea still persist among old-timers on the U.S. Gulf Coast?

In the mushroom case—monitoring. In the blowfish lore we suspect not.

You've got to monitor to understand the difference between myth and substance. And you've got to try to understand everything in order to do anything. You won't get anywhere standing on the top of your hill or sitting behind your desk telling your animals and crops what to do, forgetting for the moment about the thousands of other rebellious and independent creeping, burrowing, flying, thrusting, and twining things that surround you. You have to hark to all of them as well as forces like wind, water, and sun that you never expect to pay attention to you.

And you have to record what you learn so you can think about what it means, remember it next year, and pass it on in a comprehensible form to heirs, hands, and others—and most of all so you can use it to keep your planning vital and flexible and get better at what you do.

Don't whine when the range goes bare in January and nothing grows till June. Monitor the grass. Don't weep when starving elk bust your fences and plunder your hay. Monitor their winter range. Don't protest when your topsoil leaves for the ocean. Monitor ground cover. When old Aunt Maude left the world muttering on her deathbed that there never was a family *bundkuchen* recipe except to add this and that until it smelled right, she was telling you to . . . monitor.

PART IV
LAND PLANNING

LAND PLANNING

Many have questioned why land planning should come last in the sequence of this workbook when the basic text discusses it ahead of biological and financial planning. The answer is both practical and psychological.

The basic text introduces it first as a way to illustrate the guideline of Flexibility and the vast number of options open to holistic management. Here we discuss it last because you cannot in practice assess the possibilities without a background of financial and biological planning. More important, many people think holistic management cannot begin until they've fenced their land in some special way. In fact, whether or not you ever invest anything in fencing, you must start managing the land as you find it. A land plan is not a starting point. It is an endpoint toward which you build from your experience and success.

Conceiving the matter backwards often leads to quick disaster—yet no end of impatient people rush to borrow heavily for fences, pumps, cattle, or whatever to create a structure that proves inconvenient and cannot support itself until the whole operation matures (if ever). The social and environmental costs of an ill-wrought plan can be even worse.

Properly a land plan represents the marriage of practice and a three-part goal. Its implementation will proceed, like biological succession, as one stage makes the next one possible. Because investments in land represent long-term commitments, the Flexibility guideline from the HRM model acquires extreme importance. To quote from the textbook:

Every management decision must involve tools that offer maximum flexibility under changing conditions and commit them in ways that make flexibility possible.

Alongside financial and biological planning, land planning is a major aspect of management. It differs, however, in that it can involve very long term and occasionally irreversible commitments. Success depends on exploring all the possibilities, and trying to avoid decisions that permanently eliminate large areas of choice.

Besides flexibility the steps for developing a land plan outlined here emphasize broadness of both vision and participation. You want a land plan that goes beyond maximizing yield from a particular kind of animal. In combination with biological planning, land planning should advance all three parts of your goal and enhance as many aspects of the land's fruitfulness as possible. As well, your land plan should recognize the goals and aspirations of other people who may depend on the land in one way or another.

Especially in countries where revolution feeds on land hunger, meeting the Society and Culture guideline represents a terrific challenge. Even under less explosive conditions, however, the success of an enterprise depends as much on the enthusiasm of workers and community people as on the zeal of a manager. Thus a plan that may affect life on a piece of land for decades should include the input of everyone affected, including neighbors and communities.

Anyone can whip out a "plan" without following the steps described in this section, but no abstract blueprint for a distant future will have as much immediate *practical* use as the knowledge you'll gain from following each of the steps. In fact, your final plan will develop almost as a by-product of the process. The steps are:

1. Collecting information

2. Preparing maps and overlays

3. Creative planning:

    • Brainstorming plans based on topography alone

    • Making one plan based on existing structures

    • Designing an ideal plan based on the best ideas

4. Developing the plan gradually through annual financial planning with the HRM model so that each investment makes rather than costs money

The first section, "Mastering The Basics," will cover the first two steps. The second, "Creating Your Plan," will explain how to gather ideas and turn them into a formal plan.

MASTERING THE BASICS

The big challenge in land planning is establishing open, well-informed communication among the planners, and no degree of sensitive leadership and trusting relationships can achieve this without good information and good maps. Lacking that ballast, good discussions tend to list into arguments that cannot be resolved.

Because land planners must anticipate far into the future, they often drift dangerously into the realm of theory in the best of circumstances. Spare no pains in gathering every scrap of background material that will help you plan wisely.

Collecting Information

Approach land planning as a formal procedure in which you seek ideas from everyone who may have an interest in it or knowledge to contribute. The process of gathering information itself brings together the various parties on whom the plan ultimately depends. Though goal-setting sessions may have taken place in the past, once wrought on the land they stake out the same claim to immortality as a carved stone or a published work. People who participate become supporters—even if they say no more than "Here's where our children catch the bus" or "The best berries grow on that slope" or "Our granddad is buried under yonder tree."

If social, political, and personal tensions make the prospect of a roundtable discussion uncomfortable, consider working on better relationships before you go any further. On the other hand, the potential for destructive conflict may be less than you fear—if you keep several points in mind:

- Don't start by asking people to accept compromises. At this point you want to know their goals, desires, and worries, and what they are willing and able to do, not solutions and commitments.

- Meet people on neutral ground. There may be certain groups or individuals who wield tremendous power over the fate of the plan, but won't express themselves freely in certain contexts—especially when there are long-standing divisions of race, tribe, class, culture, or provincialism. In some situations, thanks to historic distrust, people would sooner risk cutting a landlord's fences after dark then speak up in his house in daylight to express an honest grievance. In their own councils, however, they will speak frankly.

- Allow plenty of time. How many times after an elaborate presentation have you heard a speaker say, "Any questions? Okay, since everyone agrees, we'll proceed." In most cases people need weeks to think up the key questions. Call for opinions, circulate them, call for more.

- Make a checklist of issues that need thought, and add to it. The list given here covers many that have figured in the experience of others, but each situation is unique. For each item on the list you want to know: Where on the land do we need to consider this? Who is involved? What would I like to see happen?

Checklist

— Estate planning

— Area future—roads and so forth

— Wildlife factors

— Water sources

— Weather—rain, snow, wind, cold

— Field crops and other agricultural production

__ Potential stocking rate and history

__ Other users and uses

__ Fixed unchangeable features

__ Labor—herders, hands, facilities

__ Public lands, multiple-use lands, range site information

__ Landscape goal of ranch or land unit

__ Involved people to be informed

__ Cultural aspects

__ Aesthetic aspects

__ High-density recreation/vandalism

__ Endangered species

__ Predators

__ Ownership and tenure

__ Neighboring land uses

__ Archaeological sites

__ Seasonal grazing

__ Mineral rights

Your own checklist can be much more specific, of course. Under wildlife, for instance, you might want to consider game and nongame species. Under recreation you might want to think about hunting and hiking. When you have written down people's opinions on all these issues and other local concerns and they've circulated long enough to ferment a bit, you can begin to consider specifics.

Preparing Maps and Overlays

Record your information on maps whenever appropriate, and make them available to all the people involved in planning. You cannot invest too much in this effort. Even people who work close to the land every day of their lives have trouble understanding the relationships in a large unit without maps—particularly if forests or topography prevent them from getting a broad view of it.

Try to make large-scale wall maps out of the best topographic sections you can find, and post these where people work or pass by. Aerial survey or even satellite photographs are now available for almost any place in the world. Even those who work with maps every day need time to absorb and reflect on the vast amount of information a map presents.

The untapped potential of maps in community discussions first came home to me as a teacher on the Navajo Indian Reservation. Once I put up both an aerial photo montage and a wall full of topographic maps to illustrate some tiresome point in the math book. Within a very short time finger smudges revealed a whole host of local issues I had not uncovered in years of daily contact with people. At first, people who knew nothing of contour lines, scale, symbols, or English labels regarded my displays as incomprehensible bits of abstract art—but the minute they saw the connection to the land, they learned very fast. Soon grandparents who spoke only Navajo and had never darkened the door of a school pored over them, detailing range disputes, dried-up springs, forgotten cornfields, ancient ruins, the habitat of medicinal plants, and myriad other matters that without the graphic aid of a map they could not begin to communicate to a stranger. On more than one occasion the maps were commandeered for community meetings where they perhaps cut hours off of hot discussions—though no doubt some resented how much they often revealed of an issue.

For planning purposes the master map must not show existing fences, water points, roads, and the like lest they force thinking into the old pattern. Put these features instead on an overlay to use later in testing new ideas.

EXAMPLE

Breadth and Depth in Your Thinking

You'll begin to understand why gathering information must precede drawing lines on maps as you confront decisions that can influence everything that follows—such as whether to hire herders or build fences or how you lay out croplands.

Herding vs. Fencing

Are there public opponents to fences? Are there legal restrictions on public or leased lands? Can herding ease predator problems?

What about labor costs and availability? Herding requires a good deal of skill and initial training of animals. Can you substitute dogs for people? What do you know about stock dogs, and where can you learn more?

Are there partial options involving seasonal herding on some land, skillful use of natural barriers, drift fences, movable fences, poly wire, and so forth, in combination with herding?

How about the economic question? An example in a later section on figuring the cost and sequence of development illustrates the process of thinking through a simple case to determine the herd size that will support herders, but you should gather the data as early as possible.

Cropland

Now, and probably more in the future, you may plan crop and livestock production as one unit, integrating rotations such as the corn-soy-wheat-clover sequence often seen in the Midwestern United States into the feeding of livestock.

Aside from the usual advantage of some grazing and hay production, your information gathering may turn up ways to apply animal impact, strategic feeding of winter hay, or other techniques.

And what you discover about the relationship of field size and crop sequence to weed and pest control and wildlife may well lead to a fence layout that lets you incorporate fields into the grazing plan easily.

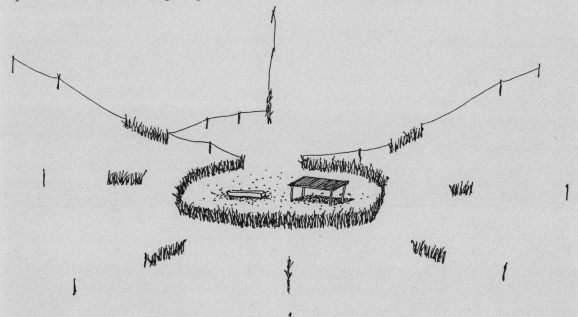

In herding situations, informal fences extending some distance from the center, whether a Navajo camp or a high-tech handling facility, can help direct traffic where a big sacrifice zone would otherwise develop.

Using Overlays

Have maps you can write on. The technology of overlays is far advanced. From an engineering supply house, art store, or blueprinter you can get several different kinds of clear plastic and both indelible and wipe-off markers in a rainbow of colors. You can also get reproductions, on white paper or film, of outlines drawn in black ink or heavy pencil.

Ideally you should have several overlays showing different considerations:

- Hunting areas

- Winter wildlife range

- Wildlife factors, roosting, mating, and nesting sites, food plants, movement routes

- Deeds, leases, and permits

- Rights-of-way

- Multiple uses

- Fire danger and prevailing wind

- Croplands and rotations with grass and legumes

- Inaccessible areas

- Existing facilities and fences

- Water sources

- Snowdrift areas

- Flood areas

- Estate plans

- Mineral rights and leases

How many overlays you have is a matter of choice. Usually, though, the information you need will fall into three or four categories. You should prepare precise information such as ownership boundaries and existing fence lines before any planning sessions. Much of the other data can be assembled on the spot by small groups armed with colored china markers—one of several marking tools that can be wiped off plastic overlays with a paper towel.

Study the maps on pages 130 and 131. Note how the fence pattern on the second map immediately suggests a cell layout—even though in this case it derives from old property boundaries that have no relationship whatever to the lay of the land. This is why you should do your creative planning on a map stripped down to basic geography.

Measuring Acreages

Throughout this book we talk about acres in one context or another. In land planning, you must constantly relate acreage to stocking rates, cell sizes, and of course the land itself. Unfortunately most people are terrible judges of area. Though 1 acre happens to be about the size of an American football field, who can easily tell how many football fields fit into an irregular 12,000-acre cell with a river and a highway through the middle?

Squares of overlay material or even thin paper divided into 10-acre squares make these measurements easy on a map. The grid shown on the next page is for the 1:24,000 scale U.S. Geological Survey "Seven Minute Series" map often used in planning. Remember: you're not buying urban lots by the square foot. A quick and dirty estimate by grid does fine.

Among other things, such grids show there is some logic after all in the English system of measurement. One-eighth mile squared is 10 acres, and 1¼ mile squared is 1,000 acres. You can take both of these units right off the scale of any map. Metric measurements don't work out quite so neatly. One-hectare squares are too small to count in many situations. A 10-hectare square is too large—and, being 316.227 meters on a side, is a pain to reckon. If you want to go metric, use 4-hectare squares, 200 meters (1/5 kilometer) on a side. Four hectares happen to be almost exactly 10

Measuring Acreage Using a Grid and Topographic Map

Making a 1,000-Acre Grid with Ruler and Map Scale

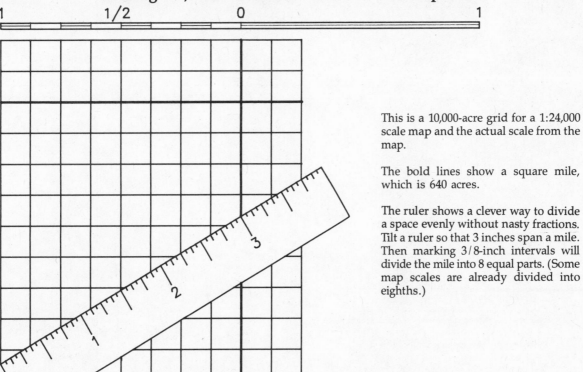

This is a 10,000-acre grid for a 1:24,000 scale map and the actual scale from the map.

The bold lines show a square mile, which is 640 acres.

The ruler shows a clever way to divide a space evenly without nasty fractions. Tilt a ruler so that 3 inches span a mile. Then marking 3/8-inch intervals will divide the mile into 8 equal parts. (Some map scales are already divided into eighths.)

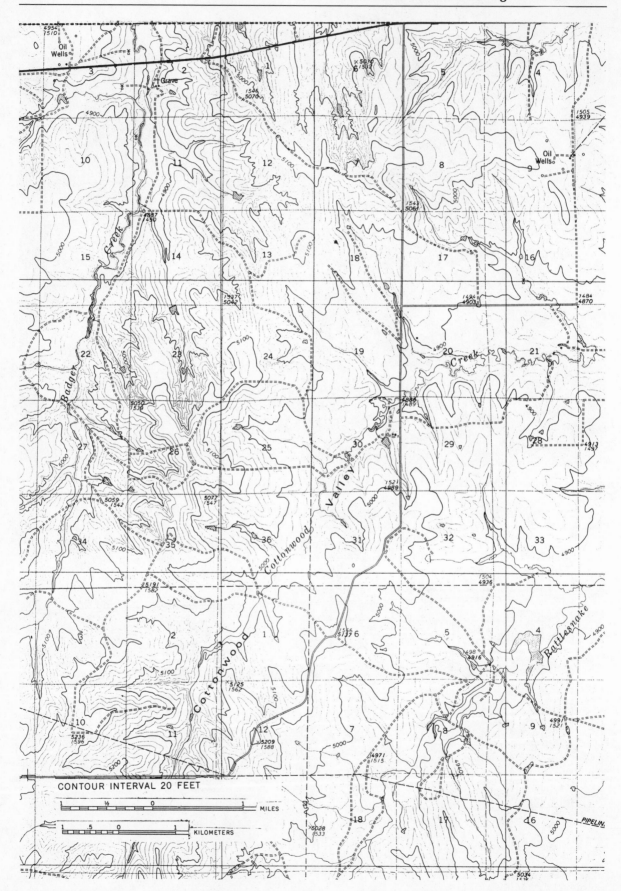

CONTOUR INTERVAL 20 FEET

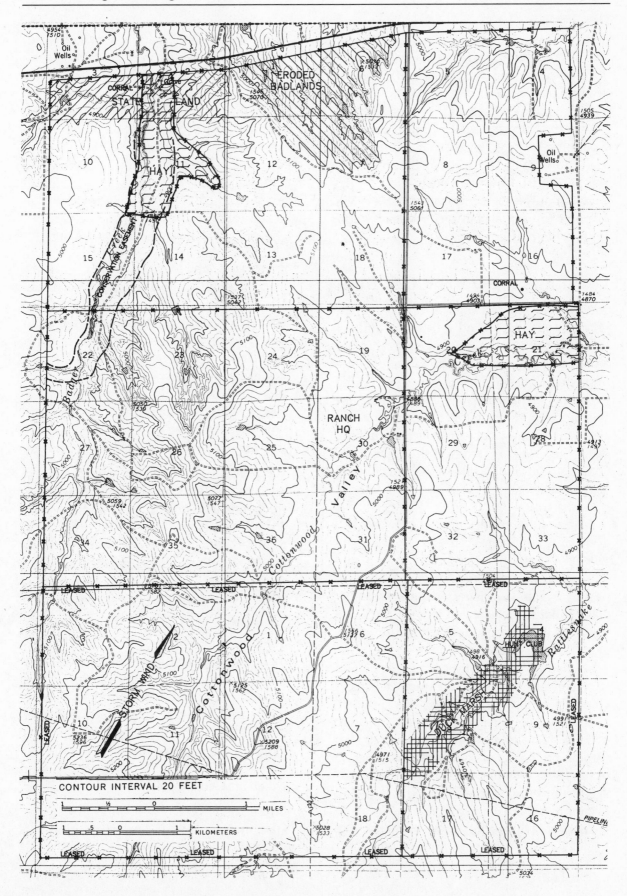

CONTOUR INTERVAL 20 FEET

acres. As you count them, 4, 8, 12, and so on, just remember that if you had stuck with rods, chains, and furlongs, you could be counting by tens.

Government offices and professional surveyors measure acreages with a planimeter, a cunning little instrument that computes the size of an area when a stylus is run around the perimeter. Cows and sheep don't worry much about accuracy, however. Unless you happen to own a planimeter already, stick with grids and guesses.

Deciding on Herd and Cell Sizes

You'll need a rough idea of how large the grazing cells will be and how many animals you might run in them before you can make any decisions on land divisions, water developments, and the like. As in many other aspects of management, you can't determine this precisely through rigorous mathematics because too many subjective factors enter in. But to get your thinking on the right track, consider your land as flat and uniform with water available anywhere (if you're willing to dig for it).

Where such conditions actually exist, a good argument can be made for a "wagon wheel" cell with water at the center and fences radiating at precise intervals. The radial design in fact has many advantages, especially in large, dry areas. Once you let go of the idea that the paddocks must be equal, symmetrical, and actually look like a wagon wheel, you can shape them to fit rough and hilly country and still have them feed to a common center.

Whether you vary the design or build something quite different, you need to start thinking in terms of area per herd, location of handling facilities, and distance to water. A circle or square representing the size of an optimum cell cut to scale from paper or overlay material will enable you to see on the map

how these relationships apply to *your* land.

Your final decisions may bear no resemblance to these "planning circles"—and in any case no cell is an immutable structure. We define it only as "a piece of subdivided land, planned as a unit." In practice, therefore, you can treat any combination of paddocks associated with any number of handling and water facilities as a cell. Nevertheless, keeping in mind the chunks of land you'll probably plan as a cell will greatly focus your thinking.

The principal variables are as follows:

- *Herd Size:* Cattle may thrive in herds of any size. No limit has been found. Nevertheless, labor, handling facilities, water, and so forth, become unwieldy after a certain point, which you must determine for yourself. If you set up a cell big enough for your optimum herd now, and better management and improving land doubles carrying capacity, you may have to run more animals at one time than you can gracefully manage just to make efficient use of forage.

- *Water Supply:* Although cattle in Hawaii can survive on dew and succulence, water is frequently a limiting factor elsewhere—and 15 gallons per day per SAU is a common requirement. If you know your water potential, herd size, and projected stocking rate, you can compute the cell size limit. In this context, distance to water also enters the equation.

- *Marginal Reaction:* In general, development cost per acre goes down as cell size increases—but figure this carefully. The same number of radial paddocks require less fencing when arranged around several centers than formed by extending fences long distances from a single center. On the other hand, supplying each center with water may cost a lot.

- *Land Boundaries:* Customs and prejudices surrounding land tenure often inhibit a flexible viewpoint, but both economics and ecology frequently justify management units quite unrelated to ownership units.

If circumstances warrant and cooperation seems possible, joint projects may make a lot of sense. Keep an open mind.

Making Planning Circles

Acres ÷ 640 acres / square mile = square miles

$\sqrt{\text{square miles}}$ = side of square cell

$\sqrt{\dfrac{\text{square miles}}{3.14}}$ = radius of a circular cell

Miles x 63,600 inches/mile ÷ map scale = inches on map (kilometers x 10,000 ÷ map scale = centimeters on map).

Hectares ÷ 100 hectares / square kilometer = square kilometers

$\sqrt{\text{square kilometers}}$ = one side of square cell

$\sqrt{\dfrac{\text{square kilometers}}{3.14}}$ = radius of circular cell

Example

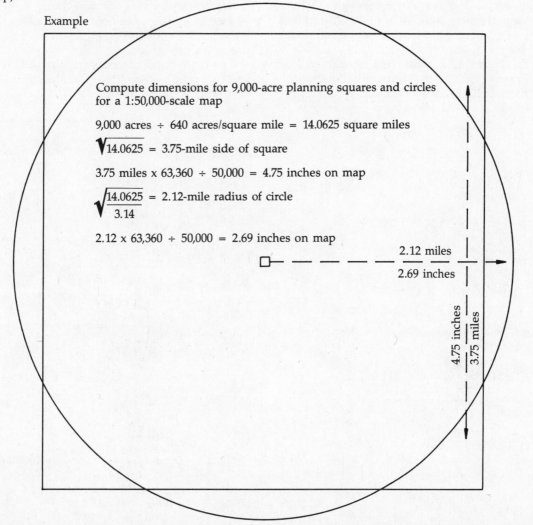

Compute dimensions for 9,000-acre planning squares and circles for a 1:50,000-scale map

9,000 acres ÷ 640 acres/square mile = 14.0625 square miles

$\sqrt{14.0625}$ = 3.75-mile side of square

3.75 miles x 63,360 ÷ 50,000 = 4.75 inches on map

$\sqrt{\dfrac{14.0625}{3.14}}$ = 2.12-mile radius of circle

2.12 x 63,360 ÷ 50,000 = 2.69 inches on map

2.12 miles
2.69 inches

4.75 inches
3.75 miles

Now for the mathematics. The formula will give you, in miles, the radius of a circle or the side of a square of a given area. For 1:24,000 scale maps you can con-vert that distance to inches by multiplying miles by 2.64. But usually it's easier just to use the miles scale on the map to draw the planning figures on paper.

EXAMPLE

Thinking Through a Cell-Size/Herd-Size Problem

Let's say you have land that has supported a stocking rate of 1:20 for many years under continuous grazing. Now you contemplate running herds of at least 600 head.

According to the chart on page 135, a 12,000-acre cell would accommodate one herd. The line at the bottom of the chart gives a radius of 2.44 miles (the greatest distance to water) for a circle that big. Could your livestock handle that?

If you think so, consider what happens when your carrying capacity increases—as you have reason to expect, judging by historical data and the performance of a neighboring ranch under holistic management.

A 50% increase in carrying capacity to 1:15 will mean 800 head. Can you handle that?

If capacity doubles to 1:10, can you increase to 1,200 head? And if so, how long will it take?

Can you supply 18,000 gallons of water daily to 1,200 head?

If you expect to double your stock, perhaps you should build two 6,000-acre cells to take your 600 head now. What will that cost?

If you build one 9,000-acre cell now, perhaps holistic management will allow a stocking rate of 1:15 now, which would accommodate your 600 head. Then an increase to 1:10 would raise your herd to 900. Is that better than 1,200?

Let's say that water limits you to 600 head, and you see no chance of developing more. If you figure you can count on a 1:15 stocking rate 9 years in 10, what options will you have when drought does strike?

This is the kind of reasoning that will lead you to a rough idea of the planning unit that best suits your conditions.

One 12,000-acre cell
600 head @ 1:20 acres
30 paddocks
Distance to water: 2.44 mi.
Fence: 89 mi. @ $1,000/mi.
One center: $5,000
Total cost: $94,000
Cost/head: $157
Cost/acre: $7.83

Two 6,000-acre cells
600 head @ 1:20 acres
30 paddocks
Distance to water: 1.73 mi.
*Fence: 74 mi. @ $1,000/mi.
Two centers: $10,000
Total cost: $84,000
Cost/head: $140
Cost/acre: $7

*Assumes that cells don't share fence.

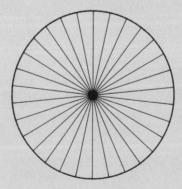

One 9,000-acre cell
600 head @ 1:15 acres
30 paddocks
Distance to water: 2.12 mi.
Fence: 77 mi. @ $1,000/mi.
One center: $5,000
Total cost: $82,000
Cost/head: $137
Cost/acre: $9.11

Grazing Cells and Herd Sizes
Possible Cell Sizes (acres)

Stocking Rates	100	200	300	400	500	600	700	800	900	1,000	2,000	3,000	4,000	5,000	6,000	7,000	8,000	9,000	10,000	11,000	12,000	13,000	14,000	15,000	16,000	17,000	18,000	19,000	20,000	21,000	22,000	23,000	24,000	25,000
1:1	100	200	300	400	500	600	700	800	900	1,000	2,000																							
1:2	50	100	150	200	250	300	350	400	450	500	1,000	1,500	2,000																					
1:3		67	100	133	167	200	233	267	300	333	667	1,000	1,333	1,667	2,000																			
1:4		50	75	100	125	150	175	200	225	250	500	750	1,000	1,250	1,500	1,750	2,000																	
1:5			60	80	100	120	140	160	180	200	400	600	800	1,000	1,200	1,400	1,600	1,800	2,000															
1:6			50	67	83	100	117	133	150	167	333	500	667	833	1,000	1,167	1,333	1,500	1,667	1,833	2,000													
1:7				57	71	86	100	114	129	143	286	429	571	714	857	1,000	1,143	1,286	1,429	1,571	1,714	1,857	2,000											
1:8				50	63	75	88	100	113	125	250	375	500	625	750	875	1,000	1,125	1,250	1,375	1,500	1,625	1,750	1,875	2,000									
1:9					56	67	78	89	100	111	222	333	444	556	667	778	889	1,000	1,111	1,222	1,333	1,444	1,556	1,667	1,778	1,889	2,000							
1:10					50	60	70	80	90	100	200	300	400	500	600	700	800	900	1,000	1,100	1,200	1,300	1,400	1,500	1,600	1,700	1,800	1,900	2,000					
1:15								53	60	67	133	200	267	333	400	467	533	600	667	733	800	867	933	1,000	1,067	1,133	1,200	1,267	1,333	1,400	1,467	1,533	1,600	1,667
1:20										50	100	150	200	250	300	350	400	450	500	550	600	650	700	750	800	850	900	950	1,000	1,050	1,100	1,150	1,200	1,250
1:25											80	120	160	200	240	280	320	360	400	440	480	520	560	600	640	680	720	760	800	840	880	920	960	1,000
1:30											67	100	133	167	200	233	267	300	333	367	400	433	467	500	533	567	600	633	667	700	733	767	800	833
1:35											57	86	114	143	171	200	229	257	286	314	343	371	400	429	457	486	514	543	571	600	629	657	686	714
1:40											50	75	100	125	150	175	200	225	250	275	300	325	350	375	400	425	450	475	500	525	550	575	600	625
1:45												67	89	111	133	156	178	200	222	244	267	289	311	333	356	378	400	422	444	467	489	511	533	556
1:50												60	80	100	120	140	160	180	200	220	240	260	280	300	320	340	360	380	400	420	440	460	480	500
1:55												55	73	91	109	127	145	164	182	200	218	236	255	273	291	309	327	345	364	382	400	418	436	455
1:60												50	67	83	100	117	133	150	167	183	200	217	233	250	267	283	300	317	333	350	367	383	400	417
1:65													62	77	92	108	123	138	154	169	185	200	215	231	246	262	277	292	308	323	338	354	369	385
1:70													57	71	86	100	114	129	143	157	171	186	200	214	229	243	257	271	286	300	314	329	343	357
1:75													53	67	80	93	107	120	133	147	160	173	187	200	213	227	240	253	267	280	293	307	320	333
1:80													50	63	75	88	100	113	125	138	150	163	175	188	200	213	225	238	250	263	275	288	300	313
1:85														59	71	82	94	106	118	129	141	153	165	176	188	200	212	224	235	247	259	271	282	294
1:90														56	67	78	89	100	111	122	133	144	156	167	178	189	200	211	222	233	244	256	267	278
1:95														53	63	74	84	95	105	116	126	137	147	158	168	179	189	200	211	221	232	242	253	263
1:100														50	60	70	80	90	100	110	120	130	140	150	160	170	180	190	200	210	220	230	240	250
1:110															55	64	73	82	91	100	109	118	127	136	145	155	164	173	182	191	200	209	218	227
1:120															50	58	67	75	83	92	100	108	117	125	133	142	150	158	167	175	183	192	200	208
1:130																54	62	69	77	85	92	100	108	115	123	131	138	146	154	162	169	177	185	192
1:140																50	57	64	71	79	86	93	100	107	114	121	129	136	143	150	157	164	171	179
1:150																	53	60	67	73	80	87	93	100	107	113	120	127	133	140	147	153	160	167
1:160																	50	56	63	69	75	81	88	94	100	106	113	119	125	131	138	144	150	156
1:170																		53	59	65	71	76	82	88	94	100	106	112	118	124	129	135	141	147
1:180																		50	56	61	67	72	78	83	89	94	100	106	111	117	122	128	133	139
1:190																			53	58	63	68	74	79	84	89	95	100	105	111	116	121	126	132
1:200																			50	55	60	65	70	75	80	85	90	95	100	105	110	115	120	125
Radius (Miles)										0.7	1.00	1.22	1.41	1.57	1.73	1.86	2.00	2.12	2.23	2.33	2.44	2.54	2.64	2.73	2.82	2.91	2.99	3.07	3.15	3.25	3.31	3.38	3.45	3.53

CREATING YOUR PLAN

When you've completed the background work—gathered information and opinions on the status and uses of the land, recorded all your planning considerations on overlays, decided on the guidelines for cell and herd size—the time has come to organize some group sessions to generate a number of very different plans.

This is a critical step and benefits from the presence of outsiders. Riding the same fences and working the same pens, corrals, and water points imparts a certain wisdom, but routine blinds the imagination. It costs nothing to reject a truly loony plan, but never seeing a new idea can be very expensive.

Participants do need to understand, though, the concept of concentrating livestock and controlling recovery times. You'll have problems working with people who oppose this idea, and it may take time, tact, and effort to get the notion across. All your other background work, the overlays, and the discussion of goals will help, however, because you can argue that planned grazing promises to solve many nongrazing problems too.

A prevalent school of thought dogmatically insists on designing fence layouts to isolate range types so that, for example, a sacaton flat or riparian area could be grazed differently from a gramma-dominated slope. However, discovery of the importance of time and advances in biological planning have reduced the importance of that concept except in special cases—such as where irrigated or riparian zones have vastly different growth rates and low paddock numbers make for relatively long grazing periods.

You may find that range types of radically different productivity will in fact justify making some paddocks larger than others. As long as you can control timing, you lose nothing by manipulating the paddock layout to ease livestock handling, game manage-ment, public access, or other land uses, to reduce predation, or to manage a host of other factors.

It may also prove helpful to show some layout possibilities to introduce the idea of handling centers designed to serve associated paddocks in a cell. (Several ideas are given later in this section.) When you do this, however, be sure to emphasize the variety of possibilities rather than any particular design. The practice of holistic management has suffered greatly from the misconception that it absolutely requires wagon-wheel cells. If you suspect that such a belief exists among your planners, make a special concerted effort to show how other designs fit certain conditions far better.

Creative planning—especially if a number of people contribute to it—will generate far more plans than you can afford to record on expensive maps or even plastic overlays. One solution is to get multiple xerox or diazo copies of good topographic maps and do the creative planning on them. Even copies, however, are not cheap when you need a lot of them. If the planners have access to a good map and overlays and can refer to it as they work, rough tracings of land boundaries and such major features as creeks and canyons on sheets of butcher paper or newsprint will work nearly as well. Such crude outline maps are quick and simple to make—just tape a master to a large window and trace over it. If your good topographical map doesn't have an opaque backing, the backlighting will allow you to trace directly from that. Otherwise you'll have to make the tracing master from tracing paper or overlay material.

Armed with a good master, overlays, a good supply of planning circle cutouts of different sizes, cheap tracings or xerox or diazo copies, pencils, and markers, proceed as follows:

Step One:

Divide your planners into groups of not more than three or four. Include husbands, wives, hands, tenants, neighbors, management club members, and youth groups, but mix and match if you can. If you don't have many people, it's better to have four working individually than two pairs. Your goal, remember, is to get a large number of plans.

If certain people are likely to dominate others, put the bosses together. If you bear ultimate responsibility for the final plan and you sense that others are likely to defer to you without discussion, consider not joining a team yourself. Almost certainly your ideas will turn up as elements in other's plans. If you have a foreman, local headman, or other leader whose authority can't be questioned without embarrassment, they might advance the work better as organizers rather than planners—answering questions of history and detail, for example.

Once you've got a multitude of plans on paper, you and others in leadership positions can contribute greatly to evaluating the results without feeling threatened by having your own plan shot down.

Step Two:

Warm up with a brainstorming exercise. The game of listing solutions to some humorous problem described on page 16 loosens up the mind and promotes healthy cooperation among the planners and competition between the teams.

Step Three:

Define the task. Explain whether you want to emphasize herding or fencing and on which parts of the land you'd do one or the other. Invite discussion and more ideas on this point.

Explain the thinking behind the herd and cell size decisions and the meaning of the cell planning circles or squares.

Ask for a specific number of paddocks or divisions. Do this in order to force the planning teams to look very hard at the topography and complications in transportation, moving, handling, and watering stock. If you have a vision of 100 paddocks, now is the moment to figure out how to create them—even if you cannot actually put in that many for decades.

Emphasize that at this stage you're not committed to using existing facilities, water points, or fences—though obviously if the country is notoriously dry, planners can't put a new well in every paddock.

Step Four:

Let the teams proceed. Most people enjoy the challenge of planning, but occasionally a group fails to take the process seriously and simply draws irrelevant lines, ignores the background information, or misinterprets a map (putting paddock lines over impassable cliffs). Stand ready to supply information or advice, but try to offer advice in the form of a question. (How will a cow get from here to there?) Stop short of outright criticism. Maintaining the spirit of free brainstorming is more important than a few lines on paper.

Step Five:

Make one plan based on existing facilities. Most land managers instinctively do this first in the belief that they can derive the most "practical" plan from the current layout, but this limits everyone's thinking to a very narrow framework. Thus it's imperative to generate some completely original plans from blank maps. On the other hand, the build-on-the-past option deserves consideration, too, and often parts of it wind up in the final plan.

Doing your creative planning from blank maps in no way implies that the old layout is useless. The point is that you may want to use it differently. If you start from a pattern of existing pastures, you'll tend to visualize those pastures as cells and think only in terms of cross-fencing them. After drawing cells on blank maps, however, you may well discover that your old fences should remain—not as cell boundaries but as cross fences in cells of a very different pattern.

Planning on a Sketched Map

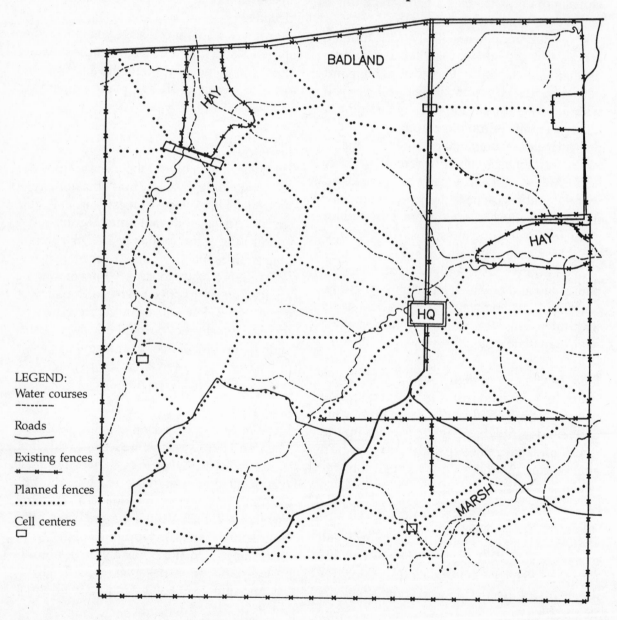

LEGEND:

Water courses
- - - - - - - -

Roads
————

Existing fences
×——×——×

Planned fences
• • • • • • • •

Cell centers
▭

Here is a land plan laid out on a sketch traced from the topographic map shown on page 130. Because the sketch shows watercourses and permanent roads, you can plan according to watershed features, even without seeing the detailed contours. Note that the plan makes use of some existing fences to divide paddocks or cells.

Step Six:
Evaluate the plans. Let the planners present their options, explaining how and why they did what they did. Don't subject anyone to direct attack. The point, as in the brainstorming exercise, is to find new ideas, not compete or keep score on bad suggestions. Do take notes, however, and use a marker or brush pen to highlight promising concepts.

After the presentations and the posting of the various plans where everybody can see them, start discussion of good elements in others' plans. When both the giver and receiver of the comment feel wise, you have a good chance of achieving consensus.

Step Seven:
Assemble the best plans. Transforming the best ideas into a workable plan may take a while. In principle, you shouldn't consider your existing facilities in the final result—though obviously you can't afford to destroy previous investments needlessly. You'll probably manage to boil down your options to two or three without much difficulty through the following review:

- Make map overlays incorporating your best ideas and superimpose them on one showing existing facilities. Can you foresee a sequence of development that will keep prior structures in use until it makes sense to replace them?

- Review water supplies. If you amalgamate herds in a drought, do you have enough flow to sustain the higher numbers in all paddocks? If not, can you afford to haul water to deficient paddocks during the occasional year you may need it?

- Go through your original checklist. Have you accounted for every item?

- Check your plans against the overlays one by one, then all together. Can you make them work?

Even when you get a couple of layouts that really look promising, do not rush to build them. Let them set a while. Let others mull them over and tinker with modifications while you work through the details of cost and construction.

Using the approach to financial planning and wealth generation outlined in earlier sections, you conceive of this task as a matter of making development of the land plan self-financing. As is frequently said in holistic management courses, investments should make money, not cost money.

Figuring Costs and Schedules

How rapidly should you carry out your land plan? The answer depends in large measure on what you determine about your situation through annual financial planning and serious consideration of the Weak Link and Marginal Reaction guidelines. Of course, non-commercial goals such as restoring a riparian zone or stabilizing a watershed might also justify investing heavily in a land plan regardless of immediate financial return.

Building cells and paddocks directly enhances energy conversion. Assuming *that* is your weak link, develop the elements of your plan that give the greatest marginal reaction in that area. The science of this is not exact, because of the number of variables and the difficulty of measuring them, but there are ways to make consistently better decisions.

First, break the entire land plan down into the smallest plausible steps, and compute the cost of each. A fencing layout might involve several stages of center construction, water development, and the fences themselves, fence by fence. There may be some items, such as a well or pipeline, that you'll have to build at once and completely in order for anything else to follow—even though you may not be able to use the full capacity of this facility for some time to come. Such cases are costly. You'll do much better if you can find a way to build incrementally so that every improvement pulls its full weight the minute it comes on line.

Second, determine a sequence of construction. All the advice in the financial planning section of this book applies here, of course, particularly these two guidelines from the HRM model, Marginal Reaction and Energy/Wealth Source and Use. The Energy/Wealth guideline will lead you to give priority to those elements of your plan that contribute most directly to the production of solar dollars. This could be the development of a water point that will increase your usable range—if you have the livestock to take advantage of it.

Far more commonly, when energy conversion is the weak link, some water is available (if inconveniently so) and the highest marginal reaction comes from raising density, increasing recovery periods, and reducing grazing periods. If you decide to do this through fencing, you'll probably find it wise to do as much as you can before investing heavily in a more efficient water system

Then the question arises over which fence to build first when you have land of different productivity and an existing layout designed according to different criteria. The following examples show how these factors affect practice.

Examples

Fences or Water?

In this example, which is used in HRM training courses, the plan calls for converting three large pastures into four radial cells. The drawing shows the existing layout in bold ink and the planned four-cell arrangement in dashes. Here are the problems:

- Assuming that developing water at each center involves fairly major capital expense, how many fences included in the new plan

could you build before having to change a water point to avoid a dry paddock?

- Assuming that, aside from the water, your centers will initially contain no expensive handling facilities and consist of only a corridor and gates, which would you build first?

- In which center would you first develop the water?

- How long would you use the existing fences (dark lines) before replacing them?

Doubtless you can think of circumstances that would dictate changing one of the water points earlier in the sequence, but you'd want to reason carefully.

Which Facilities Would You Develop First?

Existing fences ——►   Planned fences – — –◻
Existing water ▲ ■   and water

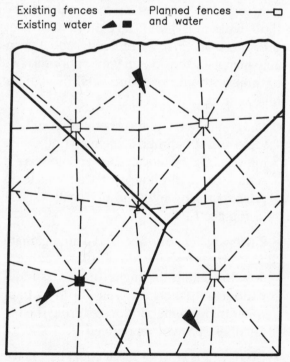

Which Fence First? The Recovery Approach

Generally you'll decide which fence to build first according to how it will reduce overgrazing on the largest number of plants. In this example, all fences cost the same and productivity is uniform across the planned paddock. The question is which two radial fences should you build first?

Choice A creates 2 small paddocks out of the 20 you have planned. They would at once get all the benefit of higher density and shorter grazing times that someday would accrue to the whole cell, though at this point there would be little impact on the rest of the cell.

Choice B, also a step toward the final plan, makes four large paddocks. It will affect the whole cell more evenly, but no part of it so intensely.

If you work out the grazing periods for 30-day and 90-day recoveries according to the formulas on page 74, the biological planning section, you'll see the advantage of Choice B:

Paddock	Min GP	Max GP
A1	2	6
A2	2	6
A3	16	48
A4	20	60
B1	10	30
B2	10	30
B3	10	30
B4	10	30

Choice A gives such unbalanced grazing periods that overgrazing will occur on 90% of the cell—both from animals staying too long in the large paddocks and from returning too soon. During rapid growth, paddock A4 will get only 20 days recovery time. And with only four paddocks you can't extend that by adding days to other paddocks without stressing both cattle and range in the smaller paddocks.

Choice B will not end overgrazing, because four paddocks do not shorten the grazing periods enough. Recovery periods will be sufficient, and 100% of the cell will benefit from the rest.

Real situations that involve range productivity, differences in fencing costs, and often many other variables are seldom so neat—as the example on page 142 shows.

Which Fence First?

Choice A

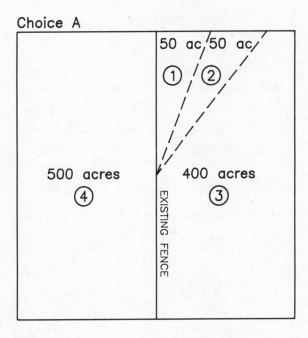

Choice B

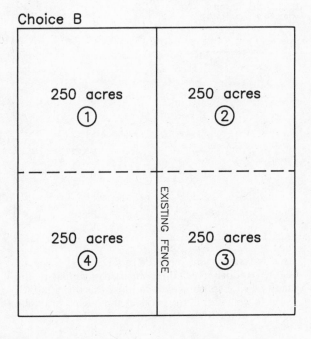

Which Fence First? The Productivity Approach

The example below includes the added variables of productivity and fencing cost. Common sense is often your best guide here, but the choice usually looks less murky if you can

Choosing Fences by AD/$

Paddock 1	Paddock 2	Paddock 3	Paddock 4
350-acre paddocks	150-acre paddocks	300-acre paddocks	200-acre paddocks
12 ADA	25 ADA	18 ADA	12 ADA
4,200 AD	3,750 AD	5,400 AD	2,400 AD
$1,500 fence cost	$2,000 fence cost	$2,300 fence cost	$1,200 fence cost
2.8 AD/$	1.875 AD/$	4.7 AD/$	2 AD/$

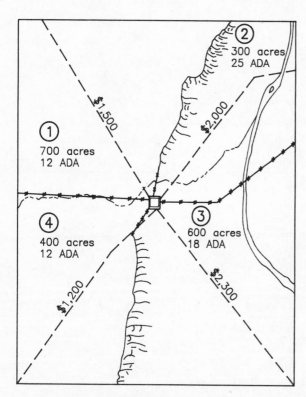

In this example you know the productivity of each existing paddock in ADA. For the sake of simplicity, the plan calls for splitting each paddock exactly in half. (If your fence will split the paddock unequally, you must use the productivity figure of the smaller piece.)

figure the productivity of new paddocks you mean to create in animal-days and divide that figure by the fencing cost. This will rate each new paddock in animal-days per dollar (AD/$)—a pretty good index of the marginal reaction of each additional fence. (Normally, you want to gain the most production for the least amount of money.)

A Composite Problem

This problem was designed from an actual case by Center for HRM instructor Roland Kroos, and illustrates the kind of choices you'll have to make in practice.

The drawings on page 143 show two possible plans for the same ranch. The broken lines show the layout of eight existing pastures being grazed as paddocks in a single cell. By subdividing them you can greatly increase density and control of time. The two plans are similar. In this case we will choose Plan B on the basis of absolute cost, though it's worth noting that the difference results from less costly water developments. The more nearly radial fence design actually requires more miles of fence.

The land varies in productivity from 10 ADA to 38 ADA, and the productivity of existing pastures varies from 1,700 AD to 28,880 AD—as you can easily reckon by multiplying AD by acres.

The Problem

Having once settled on Plan B, common sense suggests that you should start developing the fence layout by subdividing the highly productive riparian paddock 7, and all the fancy math confirms that.

First check your grazing and recovery periods. If you worked it all out for a minimum recovery period of 30 days according to the formulas in Part II on biological planning, you'd have the following grazing periods (not

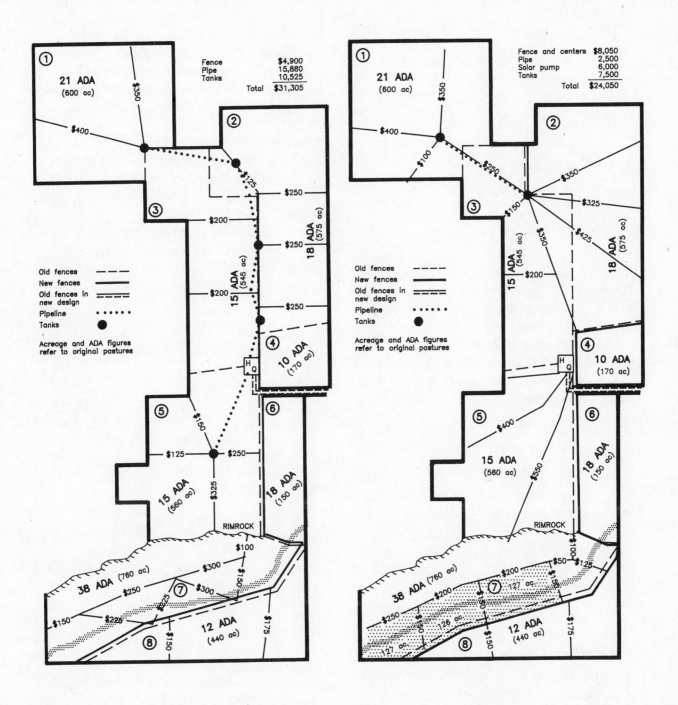

Plan A

Plan B

rounded off or adjusted) in each of the original eight paddocks:

Paddock	AD	Min GP
1	12,600	5.5
2	10,350	4.5
3	8,175	3.6
4	1,700	0.7
5	8,400	3.7
6	2,700	1.2
7	28,880	12.7
8	5,280	2.3
	78,085	

Av. AD: 9,761
Av. GP: 4.29 days

Clearly paddock 7 stands to suffer from 13 days of grazing in rapid growth, and its riparian nature will probably exacerbate the damage by assuring fast growth even during drought.

The AD/$ rule shows that fencing the shaded area gives the best marginal reaction.

At 38 ADA (it could be higher or lower than this paddock average) the shaded area rates

38 ADA x 380 = 14,440 AD

Fence cost = $250 + $200 + $200 + $500 = $800

14,440/$800 = 18.05 AD/$

This would of course only change the grazing period in paddock 7, but there it would halve it to 6.4 days on each half.

You can work out for yourself what the next division should be. One of the $150 fences crossing the shaded area looks like a good bet.

The Herding Option

As noted before, every case has its own set of variables, many of them unique. Nevertheless, consider the following example of how labor might prove more flexible than capital investment—and return more immediate benefit.

Suppose you have 36,000 well-watered but unfenced acres on which you must pay $180,000 a year. Historically this land has carried a stocking rate of 1:20. You'd like to make the mortgage payment, maintain the ranch, and net $30,000 for family living expenses in a summer of running yearlings on your grass. Perhaps you can make some extra from hunters, Christmas trees, and cross-country skiers, but most of all you want to spend the winter finishing a novel about murder and intrigue in the international commodities market without having to worry about heating the house and feeding the kids.

At a 1:20 stocking rate, using family labor, and relying on cattle prices that actually occurred in 1988 (a year most yearling operators survived) the numbers look like this:

Fixed costs	
Mortgage payment	$180,000
Maintenance	20,000
Personal needs	30,000
Total	$230,000

Animal Costs	
500-lb steer x $.95/lb =	$475
Expense of care	60
3% death loss allowance	16
Total	$551

If the steers sell at 830 lbs, the cost/lb will be $551/830 lb = $0.664/lb.

If you finance 75% of this cost at 12% for 6 months, the cost rises to $0.694/lb.

Sale and return
If feeder steers sell for $0.81/lb, you will realize $0.116/lb gross margin. That is $96.28/animal or $173,000 for the season on 1,800 head.

To cover the remaining $56,696 in overhead, you'd better negotiate the movie rights for the novel or cut a lot of Christmas trees.

Running more cattle would clearly solve your problem. You dream of four 9,000-acre cells with a total of 32 paddocks, but your plan calls for about 120 miles of fence ($150,000)—which would take both capital and time that you can't spare.

If you hired three hands at $1,500/month for 6 months to herd your yearlings, how many cattle would you have to run to pay the mortgage and break even?

Costs to be covered by yearlings:

Fixed costs	$230,000
Labor	27,000
Total	$257,000

Your break-even herd size would be:

$257,000/96.28 gross margin per head = 2,669 head

That would mean increasing the stocking rate by 50%—which might be possible given biological planning and good herding.

If the land improved enough to support a stocking rate of 1:10, then the surplus would be $89,608—plenty to underwrite fence building if you choose to do it.

Such is the flexibility offered by the herding option. But before you rush out and borrow a million plus dollars to act on such a scheme, remember that this example was created by someone doodling with a calculator, not real cattle, cowboys, and credit. It doesn't hint at the spectacular wreck that could ensue.

If, because of bad weather, faulty range assessment, or bad management, you had to sell 650-pound steers, your cost per pound would be close to $0.89 including interest—and selling at $0.81 you would lose $52 a head. That would be $93,600 on 1,800 head, but $187,200 on 3,600.

Don't double your numbers without thorough biological planning. And always keep in mind that the half million dollars and more in operating capital that the game demands would earn $50,000 to $60,000 resting safely in a bank while you wrote your novel. If you really want to own a ranch, you may have to spend 52 weeks a year at it and limit your literary concerns to whatever is stacked up on the back of the john.

Allotting Time and Money for Development

All kinds of loss and headache can result from projects that come on line too late because of some bottleneck in the work itself or the financing. A fortune can vanish with breathtaking speed if lambing pens aren't ready when lambs start dropping or a water point isn't ready when the summer gets hot.

The procedure described in Part I on financial planning shows you how to work the cash requirements of any long-term project into your financial projections. If you're careful, you should have cash on hand when you need it and know what your debt level will be at any point.

Any major construction project—such as a cell center or miles and miles of fencing—probably deserves its own column on the financial plan. It certainly requires a separate worksheet for planning the progress of the work and no doubt a lot of extra documentation on top of that.

Most planning systems include at least the following steps:

1. Break down the project into separate tasks. My father, a production engineer for a huge paper mill, used to do this on file cards so he could shuffle the chronological order later. In fact, he usually made his analysis backwards—starting with the completed work, breaking that into large categories, then subdividing each of those into smaller and smaller tasks. For example: A completed cell center might require a water system, outside fences and gates, and handling facilities, and each of these

features would be broken down into tasks and subtasks represented by separate stacks of file cards.

2. Assign a time requirement and a cost to each task. (Order pump $500—delivery time 3 weeks; advertise for crew $30—2 weeks; and so on.)

3. Arrange the tasks in chronological order. My father used to make a time line out of adding machine tape marked in 1-inch intervals (30 feet to the year, often cut to 6-foot sections to stretch out on a table). Then he would lay out his cards, working backwards from the completion date. This method shows graphically what tasks must occur simultaneously.

4. Adjust the schedule—adding slack for holidays and unexpected delays, adding labor or machinery to shorten completion times where necessary.

5. Analyze the chronology for ways to improve efficiency. Opportunities for creativity and major savings are usually enormous. They range from coordinating labor and machinery (Can the same backhoe you lease to lay pipe also be used to level the floor of the shearing shed?) to finessing transport costs. (If every pickup going to town hauls back a roll of wire, can you avoid paying a trucker to fetch the whole lot?)

6. Monitor progress. Your chronology will of course have completion dates for all the various stages, but it's easy to underestimate construction times, particularly on long projects like fencing. You'll tend to project progress from what you know of continuous work, forgetting that yellow-jacket nests, bad weather, and a myriad other things always intervene. Reduce all long tasks such as plowing, fencing, or clearing to units per day and check the rate frequently. This will give you an early warning of delays.

Layouts and Hardware

Over the years, many practitioners of holistic management have accumulated a good deal of experience in solving the technical riddles posed by the need to control the time, density, and impact of animals. New approaches turn up continually, and occasionally old ones prove faulty as evidence matures. Recorded here are some of the more durable ideas.

Paddock Layouts

Planning requires you to mark the paddock boundaries. Fencing, of course, serves this function. For herders, as one of the illustrations shows, a little fence can greatly speed a day's work. But rock cairns, flagging, or natural landmarks can also be used to designate paddocks.

Natural and ownership boundaries obviously play a major role in the design of paddocks and grazing areas, but as there is usually no need to organize them around different range types, your design can reflect any number of social, managerial, political, ecological, or aesthetic considerations.

A cell in which paddocks radiate from a common point is hard to beat for flexibility. No other design allows for such a variety of moves from one paddock to another. Nevertheless, it doesn't suit all situations—long, narrow canyons, for example. And under very boggy conditions, herds might concentrate too often near a center and overtrample the ground.

In principle, you want a design that allows maximum ease of movement from one paddock to any other paddock, minimum fencing distances, and most efficient use of water. Here are some thoughts and guidelines:

- Trails are most likely to form and be most damaging when livestock move up or down steep slopes in the narrow end of a paddock near the center. By siting the center on fairly level ground you give animals enough space to move back and forth along contours.

- Trails are less likely to form along fences that follow the worst terrain, leaving the better country open to livestock movement. As a rule of thumb, try to build your fences along ridge tops and straight down the points of ridges. Even so, the location of trails is often a mystery understood only by the animals. In very hilly country, you'd be wise to build a few of the most obvious fences first and see how the stock move for a year or so. This trial will either confirm the rest of the plan or suggest changes.

- Fences don't have to be straight, though bends do require straining posts.

- On highly productive land where distances are likely to be short and high density obligatory, larger paddocks enclosed by permanent fences can be further subdivided by temporary fences. Because high density causes rapid depletion of forage and intense fouling, livestock will concentrate on each additional section of an area encompassed by a single moving fence almost as if a second fence were moved along behind them.

Centers

In most situations, you should build cell centers before developing paddocks. If you put up paddock fences that have no function until connected to a center, livestock will never respect them. In any case, *never* introduce livestock to fences that don't carry full current.

In well-watered, highly productive pastures where distances are short, initial paddock numbers necessarily high, and existing facilities adequate, you may choose to delay building a new center. Nevertheless, if your ultimate plan calls for one, design it thoroughly right at the outset, and make sure that it will fit gracefully into your construction

Paddock Subdivided with a Movable Fence

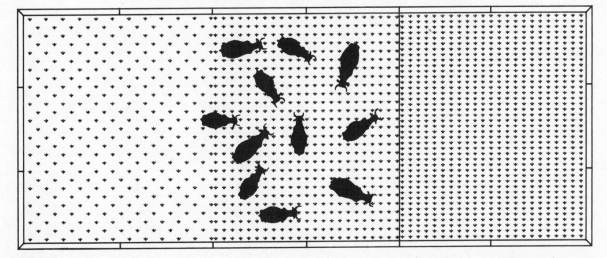

Note that no fence is required to keep stock from returning to grazed areas if grazing periods are short and density is high.

Fence Patterns

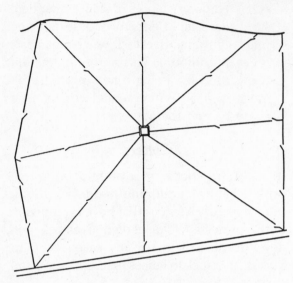

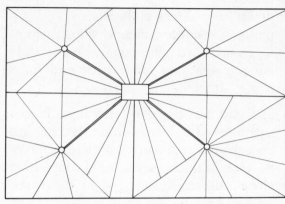

Dairy unit in center and water at satellites.

Simple radial cell with water in center. Note gates away from center.

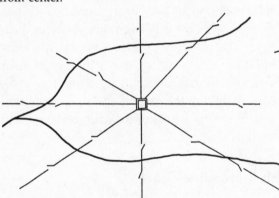

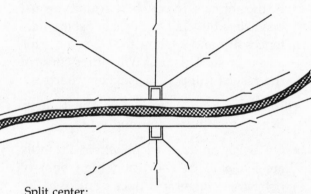

Split center:

• Full facilities would be in the half serving the largest area.

• Half centers can be set well back from river or road.

• 1-, 2-, 3-, or 4-way splits are possible.

Dry center—water available in all paddocks.

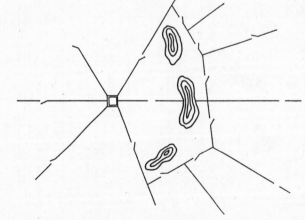

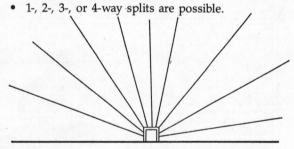

Radial fences in a fan design (takes more fencing than wheel layout).

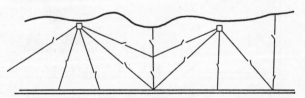

Fencing to avoid problem piece of land near center.

Several fans used in a narrow area.

Fence Patterns

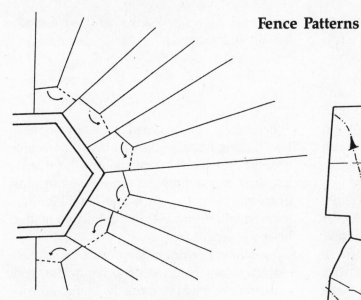

Alternate radials into a small center.

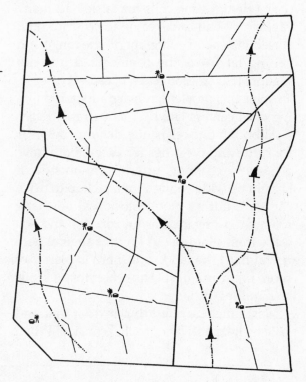

Three cells without radiating fences.

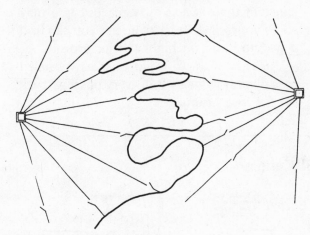

Two cells divided by rimrock.

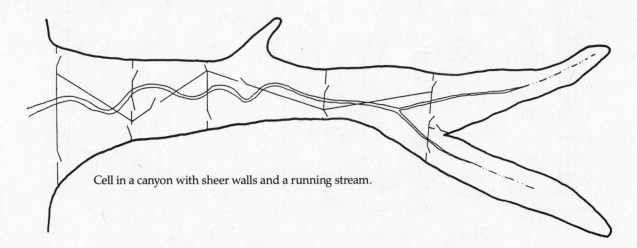

Cell in a canyon with sheer walls and a running stream.

sequence. Few rules govern center design, though long experience has uncovered a few ideas that will help you avoid the most common problems (see pages 148 to 149).

The first grazing cells for cattle had many problems. The cattle came in to drink and stayed instead of going out on the range and getting fat. Also, the center filled up with manure that smelled bad and caused disease. To solve this problem, ranchers started building cells centers like the ones drawn here.

Although one cell center diagram below is for cattle and the other for sheep, both have a narrow "corridor" around the outside. It should be wide enough for a pickup to drive through, but no more. Stock will not hang around too long in a narrow corridor, and will leave their manure out on the range, where it will help the land, but there is plenty of room in the middle of the center for all kinds of handling facilities.

Most ranchers build the corridor first and little by little add fancy facilities inside. When

you build, *always* look ahead and leave space for all future needs.

Siting

When centers enjoy broad, level approaches, less trailing results—though if you plan one around an existing water point where trailing was not a problem under continuous grazing, it won't become one now. Planned grazing will improve the situation no matter how the fences run.

Your next considerations must be access, water availability, construction obstacles, and proximity to roads or hazards. Many old corrals were built in potentially boggy areas because of their access to water, but you may spare yourself decades of grief if you can find a way to build on higher, drier ground.

You may want to build near a well-used road—but how near? Will every passing truck startle your animals?

Sample Cell Centers

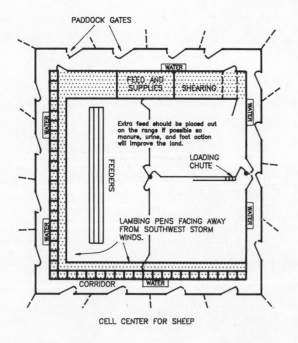

CELL CENTER FOR SHEEP

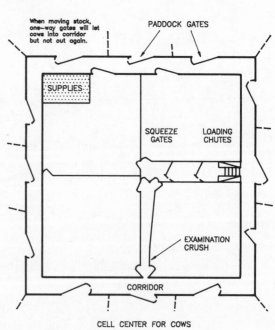

CELL CENTER FOR COWS

Corridors

Ideally you want cattle out on the range, not lounging around piling up dung and fly bait in the center. Yet you also want space in the center for handling animals, storing supplies, and corralling riding horses, bulls, and sick stock. A narrow corridor built around a central area will solve both problems. The corridor communicates with all paddocks served by the center and can contain water facilities. Animals that come there to drink will have no incentive to loiter in the confined space and chew cud, but will move back out on the range where their dung is an asset. Meanwhile, you can enjoy a large clean central working space. The corridor might even surround an entire homestead.

Generally you'll figure the size of the central space according to the facilities you want to put there, but it's easier to be generous at the outset than to expand later. Five thousand square yards will easily accommodate the needs of a herd of 500 to 1,000 cattle that need daily handling.

Plan the center for all the paddocks, herds, and facilities you might ever want to handle—even if you initially build only the corridor around a large empty space. The corridor should be wide enough to drive through, but 15 feet will accommodate herds of almost any size and the benefits disappear entirely at about 30 feet.

Many people worry about damage should two bulls fight in a narrow corridor, but this occurs almost exclusively in wide ones. In a restricted space, love triangles are less likely to form and seldom simmer long enough to explode. Even jealous contenders prefer to challenge where a retreat, if necessary, can be conducted with dignity.

The perimeter should be long enough to accommodate plenty of gates. For commercial production you'll probably look forward to 30 or more paddocks at some point.

When you've planned your center on paper, lay it out on the ground with pegs and twine or powdered lime and test it as far as possible. If vehicles will have to pass through it or turn around to back up to a loading ramp, be sure they can do it before you start digging the postholes.

Remember that the part of the center facing the portion of the cell with the greatest grazing area will tend to have the most fences coming in to gates around the perimeter. Lay out radial fence lines from the actual gates rather than from a common peg in the center—otherwise they'll all be bunched together.

Water

Unless a winding stream or other convenient water source supplies all your paddocks, you'll want to have water at the center. There are more combinations of solar, wind, and gas-powered pumps, rams, siphons, and pipes to help deliver it than this book can describe, and the technology keeps improving. Even if you don't have water where you want to put a center, don't abandon the idea without an up-to-date engineering study. Keep these principles in mind:

- You must have a flow such that livestock never have to wait. Otherwise you'll have animals packed into the corridors sniffing at drips. The dominant individuals will develop a persistent habit of driving others away. If all your livestock can trust the supply, they'll only come to the center when genuinely thirsty and then leave. This means you'll have to engineer a sufficient direct flow or make sure that storage capacity is big enough to handle peak demand. Where water sources are poor, it can pay to plan cells and then pipe small amounts of water from several minor sources to one center. This way you can supply a herd big enough to do the land some good.

Water Seep for Wildlife

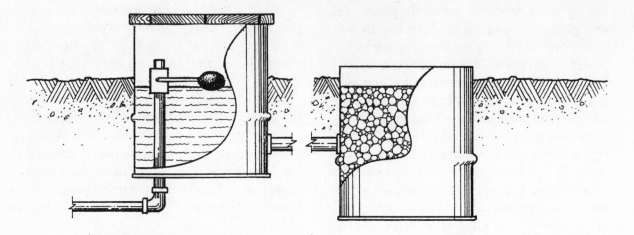

Water level in the left barrel half will maintain level in the sand and gravel on the right.

- The concentration of animals at the center can create enough dust to affect large open tanks. Long narrow troughs, served by a rapid flow, do better and are easier to clean, especially if the ends slope. (Calves may get pushed over and drowned in troughs over 9 inches wide, so bars should be welded across the top.)

- Consider building ramps or steps that allow birds and other small animals to drink without falling in and drowning. Not only will you keep the water unpolluted but you'll be helping to build up complex communities.

- For the many shy creatures that can't or won't brave cell center fences to get water, seeps or pools are easy to construct. The illustration above shows one made with a regular float valve set in half a steel drum. This device functions like the tank on a toilet. The flow runs from the covered tank into the other half of the drum buried to ground level and filled with gravel and sand. Wild things can always scrape a bit

to find a drink. An old trough or plastic-lined depression would provide an even better seep than half a drum.

Fencing to Protect a Tank or Small Pond

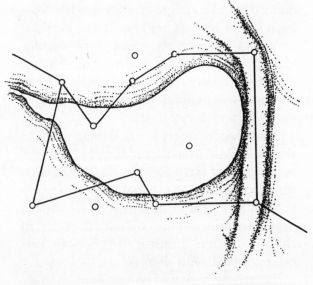

Moving wire to allow access to water at different points allows management of shore communities while preventing animals from standing in pond and fouling it.

- Dams, sloughs, and other catchments present a special challenge—especially if you use them to supply several paddocks so that stock have access to them over longer periods. Consider fencing them so that animals can't stand in them, or wallow and foul them. You can drive several permanent posts that will allow you to change the access area occasionally or extend it when the level drops.

Fences

Fencing technology is evolving so fast that any details offered here would soon be out of date. For some years now the solution most cost-efficient, versatile, and least damaging to wildlife has been the free-running slick wire fences developed in Africa and electrified by high-voltage pulses from New Zealand energizers. Variations such as poly wire that can be unrolled quickly for temporary driveways and barriers, rolling fence posts, and other innovations are continually increasing your options for creative uses.

How Many Wires?

Most people start out using far more wire than necessary, much to the delight of the fencing industry. On dry soil where grounding is poor, two wires, one hot and one grounded, will contain almost any domestic livestock—if they're not seriously stressed and if they're well-trained to respect the fence. Where grounding is good, one wire will suffice.

Frequently you can tell the sequence of fence building on a ranch by counting the strands of wire. The early fences have three, four, or more strands and closely set posts. Eventually confidence grows to the point where the most recent fences have two strands and posts set so far apart that the wires droop between them.

Spacing the Wires

No ideal spacing has yet been determined. Most reasonable ideas work, so the question is not critical. A common arrangement for cattle is two wires 3 inches apart and between 24 and 30 inches above the ground. The top wire is hot and the bottom one grounded. For smaller stock the bottom wire should be lower. You should consider having three wires in the case of mixed herds of cattle and sheep.

Contrary to a lot of conventional wisdom, the same fencing that divides paddocks will work on the cell perimeter. Except for paddocks actually containing livestock, power will be switched off in most fences, including most of the perimeter fences, most of the time.

Posts

Almost any post can be used—wood, steel, or fiberglass. Usually cost and availability will make this decision for you. Poor-quality fiberglass posts will split out, decay in sunlight, or be sawn apart by the wire vibrating in the wind, so ask for references.

Most fiberglass posts and the lighter wooden posts will slide up and down in their holes. This, however, seldom causes any problem. There are many cases of deer learning to lift up fences and pass under. The posts simply drop back into place.

You can also use growing trees, but use caution if the species has potential timber value. Theoretically you can attach an insulator to any tree without harm, but there's a good chance you'll damage or distort at least some trees or leave wires or fittings grown into the bark. This mistake could one day wreck a saw or kill a logger or mill hand. If the wire bends around the tree, you can often run it through plastic insulator hose. The tension will hold it in place without any need to risk the tree at all.

Attaching the Wires

The wires must run free, and the attachment must be strong, durable, and easy to install. New devices turn up every day.

The wires must be able to slide unrestricted. Otherwise any shock or expansion and contraction of the wire will concentrate its force on a single post. A free wire distributes the force through the whole fence to end strainers designed to take it.

Avoid steel-to-steel contact at the attachment point because the galvanized surface of the wire quickly wears through and rust sets in. Apparently some plastic connectors collect dust and grit that also cause this problem.

Beware of any connectors of weak design, especially plastics that become brittle in sunlight. Fiberglass and treated wooden posts threaded on the wire work well, but you'll have to cut the wire to replace damaged posts.

Straining Posts

At each end of the fence you'll need a straining post. And at various points in the length of the fence you'll want simpler posts.

The design shown below is a bit simpler and quicker to build than the conventional H-strainer used in barbed wire fences. Most experts recommend a diagonal of 8 to 10 feet. Set the post deep. There should be as much below ground as above. The pad at the end of the diagonal can be rock or treated wood.

Generally a single stout wooden post can handle strain in the middle of a fence, because the pull on opposite sides cancels it out. Two electric wires only add up to about 400 pounds anyway.

Grounding

One of the most common faults with electric fencing is poor grounding, especially in dry climates. If your soil and air are dry, you may well need three or four 6-foot steel rods driven in the ground and attached in series. These are connected to the charger and the ground wire in all fences. When the charger is located at the center, the power is often distributed to radial fences through wire strung around on posts above the paddock gates. If the ground wire is attached above

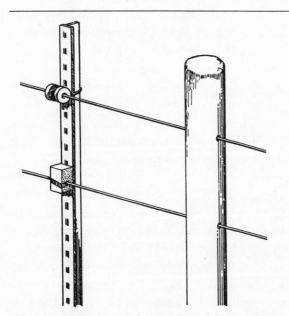

Wire attachments must be strong and free running.

A strong, simple straining post.

the hot wire, it also serves as a lightning deflector.

Gates

Although stock move a lot, most gates will only be opened and closed about a dozen times a year unless they're on well-used roads. But for the sake of flexibility you can't have enough gates. You absolutely must be able to move livestock from one paddock to the next at several points other than the center. Otherwise you'll train them to a destructive routine or sacrifice a piece of ground.

That said, any kind of gate will work. The two-strand fence doesn't require anything elaborate. Wherever your ground is level enough to allow a good distance between poles, you can push the wire to the ground with your foot. A couple of hooks buried in the ground will hold it down and horses, trucks, or herds can pass over.

Gully and Stream Crossings

The electric fence handles these easily. The wires pass straight over and have alternating hot and cold wires or chains dangling down into the gap. This barrier prevents stock from moving past but allows water and debris to wash through without damage.

To keep the drop wires from shorting, run the fence wires side by side on opposite sides of wooden posts instead of one above the other. Wires or chains hung every foot or so on opposite wires will make an effective barrier as shown on page 156.

Educating Livestock

Animals usually learn quickly to respect electric fence. Occasionally, however, it may help to tempt a particularly unruly lot to lick aluminum cans smeared with molasses and clamped to the hot wire. In dry climates where grounding is poor, this method works best on a wet day or where the cans hang so

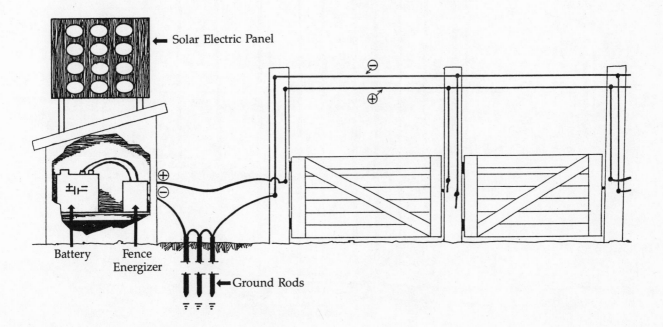

General wiring pattern for fences radiating from a center.

the animals are likely to touch the ground wire in reaching for them.

Many people who annually acquire great numbers of untrained stock like to start them out in a small "training paddock" at enough density to make sure that a good number of animals get zapped. Try adding an extra wire or two in the fence and hosing down the ground to increase the likelihood of a powerful shock. In any case, retaining a few experienced lead animals each year greatly speeds the education of the rest.

Unshorn angora goats and sheep present a special challenge. I've seen an especially wild herd of angoras ignore the molasses trick for days because they had never tasted any before. Ultimately they learned that molasses stings but fences don't. They didn't get the

point until a good rain enhanced the conductivity of both the ground and their fleece.

Training stock to move to a signal such as a whistle, bell, gong, or horn also adds immensely to the ease of handling animals. Not only does it make your fencing more effective and secure, but it's also useful for concentrating livestock to produce herd effect.

A good many experienced cowboys will swear this is nonsense. Don't believe them. There are cases of small children on foot moving fairly large herds of bison that have learned the meaning of a whistle. On other ranches, however, hands have revolted and quit because they spent all their time in the saddle gathering and driving cattle.

Training takes a little patience and a little art. Here are the ground rules:

Electric Fence Crossing a Stream

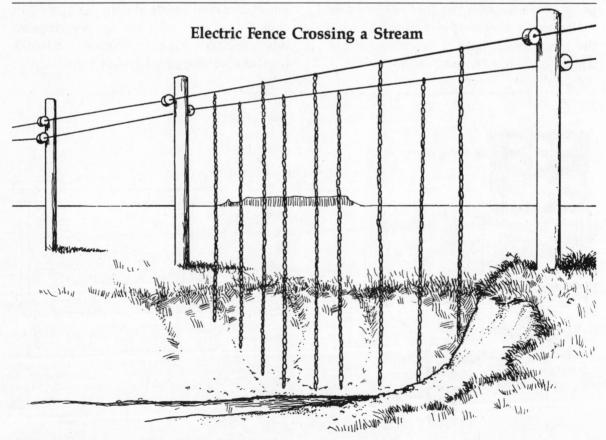

Chains hanging from parallel strands overhead create a hot barrier.

- Choose a unique signal that the animals only hear when you mean to move them. Training to a pickup horn will cause you no end of problems.

- Always associate the sound with a reward such as hay or supplement cubes. Eventually animals will find sufficient reward in a new paddock, but make the message absolutely clear while you're training them.

- Don't mix reward and punishment. If you drive them while blowing a whistle—even if you drive them toward a reward—they won't get the point.

- Mix older, trained animals with newcomers until the message gets across.

- Start the training in smaller paddocks where the animals can see you and each other.

EPILOGUE

In several places I've used the metaphor of the artist standing before a blank sheet in order to inject a sense of drama and romance into tasks that will probably keep your hand on the calculator and your nose in the paper longer than you ever thought you could bear. Certainly holistic management requires a lot of organizing, planning and monitoring. But painters and poets don't really enjoy much drama and romance in their day-to-day work. Most grind out their lives in risky, low-status toil, and they die broke far more often than ranchers and farmers do.

Nevertheless, the comparison holds in other ways. In the blood of all who take responsibility for blank pages or pieces of land runs the desire to bring forth from them something altogether new and of lasting value.

Who knows why? A hundred times you've heard faded men in faded denim say "I don't know why I stay in this business"—though you know that leaving the bone-grinding labor on their land would kill them in no time at all.

Because you do know why, you'll compose a land plan worthy of your art that will sustain you and others who live from it. You will create it from a fresh map, unencumbered by the roads, fences, ditches, and other clutter you inherited or built without forethought.

Your land will become what it ought to become.

We've already hinted at the material benefits of planning. Too many people forfeit these benefits by clinging without reflection to what always was. A little bit of patient planning may return hundreds of thousands of dollars over the years in savings and increased income from fewer roads, water points, and fences, better livestock handling, greater wildlife productivity, reduced losses from weather and predation, and a host of other factors.

There is a bigger reason, however.

While theologians debate the immortality of the soul, you can walk around and get your boots muddy in the ages of the earth, which compared to your short life might as well be immortal.

Then again, you aren't sure, are you?

No other living planet has yet been found in space, and ours is very small. We tax it sorely with our bombs, wars, fumes, and fires, our cutting down and building up, our teeming cities and plundered fields, our grasping and our greed.

And that is why you will sit down with your planning sheets, your computers, and your maps, and do your work—so that when the paintings in their galleries and the poems on their shelves have gone to dust, the earth, your piece of land, will abide.

APPENDIX A
Financial Planning Forms

The financial planning process described in Part I reflects the accumulated knowledge of several decades. Lessons learned in practice have enabled the Center for Holistic Resource Management to refine and improve the various planning steps and will continue to do so.

To ensure that HRM practitioners are kept current, the Center periodically updates its "Comprehensive Guide to Financial Plan-

ning." You can obtain this guide from the Center for Holistic Resource Management, P.O. Box 7128, Albuquerque, NM 87194.

The financial planning forms you'll need—Annual Income and Expense Plan, Worksheet, Livestock Production Plan, Control Sheet—are reproduced here (not actual size) and are also available from the Center.

ANNUAL INCOME AND EXPENSE PLAN

PLAN																						
— ACTUAL																						
DIFFERENCE																						
CUMULATIVE DIFFERENCE TO DATE																						
PLAN																						
— ACTUAL																						
DIFFERENCE																						
CUMULATIVE DIFFERENCE TO DATE																						
PLAN																						
— ACTUAL																						
DIFFERENCE																						
CUMULATIVE DIFFERENCE TO DATE																						
PLAN																						
— ACTUAL																						
DIFFERENCE																						
CUMULATIVE DIFFERENCE TO DATE																						
PLAN																						
— ACTUAL																						
DIFFERENCE																						
CUMULATIVE DIFFERENCE TO DATE																						
PLAN																						
— ACTUAL																						
DIFFERENCE																						
CUMULATIVE DIFFERENCE TO DATE																						
PLAN																						
— ACTUAL																						
DIFFERENCE																						
CUMULATIVE DIFFERENCE TO DATE																						
PLAN																						
— ACTUAL																						
DIFFERENCE																						
CUMULATIVE DIFFERENCE TO DATE																						
PLAN																						
— ACTUAL																						
DIFFERENCE																						
CUMULATIVE DIFFERENCE TO DATE																						
PLAN																						
— ACTUAL																						
DIFFERENCE																						
CUMULATIVE DIFFERENCE TO DATE																						
PLAN																						
— ACTUAL																						
DIFFERENCE																						
CUMULATIVE DIFFERENCE TO DATE																						
PLAN																						
— ACTUAL																						
DIFFERENCE																						
CUMULATIVE DIFFERENCE TO DATE																						
TOTALS																						

WORKSHEET

CENTER FOR HOLISTIC RESOURCE MANAGEMENT

Date _____

WORK SHEET NO. _____

Detail	January	February	March	April	May	June	July	August	September	October	November	December	Total
Total													

LIVESTOCK PRODUCTION PLAN

CENTER FOR HOLISTIC RESOURCE MANAGEMENT

RANCH _____

ENTERPRISE _____

REMARKS _____

YEAR OR MONTHS .... YEAR OR MONTHS .... YEAR OR MONTHS .... DATE OF PLAN ....

A CLASS OF STOCK (Cows, Heifers, Calves, Bulls, etc.)	B % EST BIRTH	C OPEN NO.	D AGE	E BIRTHS	F MONTH	G BUY	H MONTH	CLASS TRANSFERS IN / OUT	I DEATH	J %	K SALE	L MONTH	M CLOSE AND OPEN NO.	N AGE
1.								1						
2.								2						
3.								3						
4.								4						
5.								5						
6.								6						
7.								7						
8.								8						
9.								9						
10.								10						
11.								11						
12.								12						
13. TOTAL HEAD														

NOTE: RECORD YOUR ESTIMATES OF PERCENTAGE ACTUAL BIRTHS FOR THE VARIOUS AGE CLASSES OF BRED FEMALES IN THEIR ROWS IN COLUMN (B) ABOVE.

ANALYSIS OF PLANNED SALES AND PURCHASES

14. CLASS OF STOCK			
15. NUMBER SOLD/MONTH OF SALE			
16.			
17. AVERAGE LIVE WEIGHT			
18. MEAT PRICE PER LB.			
19. INCOME PER ANIMAL			
20. WOOL/HAIR WEIGHT/MONTH OF SALE			
21. WOOL/HAIR PRICE/MONTH OF SALE			
22. WOOL/HAIR INCOME			
23. PLANNED GROSS INCOME			
24. NUMBER PLANNED TO BUY			
25. ESTIMATED PRICE/ANIMAL			
26. TOTAL COST AND MONTH			

Center for Holistic Resource Management

CONTROL SHEET

Name: _____ Date: _____ Sheet #: _____

Plan Column #	Amount Adverse to Date	Cause of Deviation from Plan	Proposed Action to Return to Plan	ACT

APPENDIX B
Biological Planning Forms

The *Aide Memoire* described in Part II is continually updated as experience gained worldwide calls for revisions in or additions to the various steps. Some of these changes also dictate revisions in the Biological Plan and Control Chart that accompanies the *Aide Memoire*.

Check periodically with the Center for Holistic Resource Management to ensure that you have the most up-to-date *Aide Memoire* and chart. The Center encourages you to share any experience you might encounter that isn't covered in the *Aide Memoire* so it can be incorporated in future editions. Write to: Center for Holistic Resource Management, P.O. Box 7128, Albuquerque, NM 87194.

The Biological Plan and Control Chart reproduced here (in reduced form) can be ordered from the Center. The Worksheet shown in Appendix A (page 161) is also used in biological planning and can be ordered from the Center as well.

BIOLOGICAL PLAN & CONTROL CHART
(Livestock/Wildlife/Other Uses)

CENTER FOR HOLISTIC RESOURCE MANAGEMENT

YEAR 19 _____

GRAZING CELL _____

NORTHERN HEMISPHERE CHART

COLUMNS

1	2	3			4	5	6	7
PREVIOUS NON GROWTH ADA YIELD	ESTIMATED RELATIVE PADDOCK QUALITY RATING	PADDOCKS			MINIMUM MAXIMUM GUIDELINES	ESTIMATED ADA AVAILABLE (NON-GROWING)	ESTIMATED ADs AVAILABLE (NON-GROWING)	PLANNED DEMAND ADs OR ADA (NON-GROWING)
		No.	SIZE	PD #				

Months columns: JANUARY, FEBRUARY, MARCH, APRIL, MAY, JUNE, JULY, AUGUST, SEPTEMBER, OCTOBER, NOVEMBER, DECEMBER

Each month subdivided into: No. | AVG WEIGHT | % UNIT

Row labels:

21 RAINFALL
22 SNOW
23 GROWTH RATE (F/S/D)
24 SUPPLEMENT TYPE AND AMOUNT
25 NUMBER/SIZE OF HERDS
26 PADDOCKS AVAILABLE
27 SHORTEST/LONGEST RECOVERY DESIRED
28 AMGP / AMGP
29 ANIMAL UNITS (OR SAU S)
30 TYPE OF ANIMALS
31
32
33
34
35
36 CELL SIZE

STOCKING RATE: GROWING SEASON _____ STOCKING RATE: NON GROWING SEASON _____

MORTALITY and CAUSES: _____

REMARKS: _____

CONCEPTION RATE: _____ WEANING RATE: _____

A. Estimated Total ADs (Livestock/Wildlife): _____

B. Estimated Days of Non Growth: _____

C. Days of Drought Reserve Required: _____

D. Total Days Grazing Required: _____

E. Estimated Carrying Capacity (SAUs): _____

SUMMARY LIVESTOCK RESULTS

Calving / Lambing / Kidding	_____ %
Avg Weaner Weight / Age	_____ / _____ Mths.
Daily Avg Weight Gains	_____ Lbs.
Total Yield Per Acre	_____ Lbs.

AVG ANNUAL PRECIPITATION: _____

SEASON TOTAL PRECIPITATION: _____

APPENDIX C
Biological Monitoring Forms

The biological monitoring process described in Part III has undergone many years of refinement. Practical experience will continue to lead to further innovations. You can obtain updated versions of the "Guide to Biological Monitoring" from the Center for Holistic Resource Management, P.O. Box 7128, Albuquerque, NM 87194. The biological monitoring forms reproduced here—the Monitoring Data Form (two sides) and the Monitoring Summary Form—are also available from the Center. The Field Instruction Sheet, also reproduced here, provides a summary of the monitoring process.

Center for Holistic Resource Management

MONITORING DATA FORM

Ranch: _____ Cell: _____ Xsect. nos. _____ Photo nos. _____ Examiner: _____ Date: _____ Date HRM _____

Nrst. Plant & Point Hits		Dot Tally	Total		Dist.	C/W	Age	Form	Specie	Dist.	C/W	Age	Form	Specie	Dist.	C/W	Age	Form	Specie	Notes
Cover	bare			1																
	litter 1			2																
	litter 2			3																
	rock			4																
	basal			5																
Canopy	grass			6																
	forb			7																
	brush			8																
	tree			9																
Capping	mature			10																
	immature			11																
	recent			12																
	broken			13																
	"covered"			14																
Living Organisms	insect			15																
	bird			16																
	small animal			17																
	large animal			18																
Plant Type	grass			19																
	forb			20																
	brush			21																
	tree			22																
Habitat	dry			23																
	middle			24																
	wet			25																
Erosion	soil			26																
	litter			27																
	pedestal			28																
	flow			29																
	rill			30																
	gully			31																
	Total			32																
				33																
				100																

GROUND ATTABL GROUND ATTABL ↓ ↑

Distance	C/W	Age	Form			

Center for Holistic Resource Management

Species			Notes
	Name	#,%	
Grasses			
Forbs			
Brush			
Tree			

Center for Holistic Resource Management

MONITORING SUMMARY FORM

1	Ranch:		Cell:	Xsect. nos.		Photo nos.		Examiner:		Dates:	

2 DATA SUMMARY	% Cover	Canopy	Capping	Living Org.	Plant Type	Habitat	Erosion	Avg. Distance	AGE	FORM	Species
	bare	G-	M-	I-	G-	Dry-	S-		S-	N-	List Separately by Plant type groupings
	litter 1	F-	I-	B-	F-	Mid-	L-		Y-	OR-	
	litter 2	B-	R-	S-	B-	Wet-	P-		M-	OG-	
	rock	T-	B-	L-	T-		F-	cool-	D-	OB-	
	basal		C-				R-	warm-	R	D	
							G-				

3 Production, Land Description, and Quality of Life Goals:

4 Ecosystem Blocks: Water Cycle; Mineral Cycle; Succession; Energy Flow

5 Tools: Money/Labor (Rest Fire Grazing Animal-Impact Living Organisms Technology) Human Creativity

6 Guidelines: Testing:

Whole Ecosystem   Weak Link   Marginal Reaction/Gross Margin   Energy/Wealth   Society & Culture

Management: Time, Stock Density, Herd Effect, Population Management, Burning, Flexibility, Biological PMCR, Organization/Personal Growth, $ PMCR

Field Instruction Sheet

Step 1
Carefully review the detailed description of monitoring procedures in the previous section.

Step 2
After locating the permanent starting point for each transect, make a vertical photo at a marked point along the transect and another general photo in the direction of the transect.

Step 3
Record data:

1. Enter the transect identification information at the top of the form.

2. Randomly throw the dart and record the point hit in the appropriate cover row.

3. If the point hit was basal plant cover, record a zero in the Distance column.

4. Check for canopy cover above the point hit, using the plumb bob if necessary. If canopy directly covers the point hit, record the canopy type in the proper row.

5. Record the degree of soil capping within a 6-inch radius of the point hit.

6. Record a dot in the Living Organisms section if you see any animal or insect sign within 6 inches of the point hit.

7. Record the type of perennial plant closest to the point hit. Measure and record the distance now if the closest plant is not obvious.

8. Record the habitat type of the nearest perennial plant in the correct habitat row. Use "middle" for anything except obviously wet or dry.

9. Measure the distance from the point hit to the nearest perrenial plant. Make the measurement to the basal crown of this plant. Record the figure in the Distance column. If the point hit was basal, all further information except canopy type will concern the plant contacted by the point hit. Be sure to acknowledge the presence of annuals, if any, with a check mark.

10. Record the growth season for the nearest perennial plant ("C" for cool season and "W" for warm season).

11. Record the age class of the nearest perennial plant (or the plant struck by a basal hit). Use "S" for seedling, "Y" for young, "M" for mature, "D" for decadent or dying, and "R" for resprout.

12. Record the growth form of the nearest perennial plant (or basal hit plant). Use "N" for normal, "Or" for overrested, "Og" for overgrazed, "Ob" for overbrowsed, and "D" for plants dying from other causes.

13. Record the species, using an abbreviation, if you know the name.

14. Repeat items 2 to 13 until you have completed 33 point hit measurements.

15. Fill in the first subcolumn of the Erosion section (using the Erosion Condition Guide) for the transect so far. Total this column.

16. Repeat items 2 to 13 until 66 point hits have been recorded.

17. Fill in the first subcolumn of the Erosion section for the second third of the transect. Total this column.

18. Repeat items 2 to 13 until 100 point hits have been recorded.

Step 4
Combine the data.

1. Average the rows in the Erosion section and put these in the total column to the right. If you think the erosion conditions in the last third of the transect differ greatly from those in the first two-thirds, make a third erosion assessment in the last subcolumn and average all three.

2. Total all dot count information in the total column to the left of the Dot Tally.

3. Subtotal the information for each group of 33 points (34 in the last group) at the bottom of the page.

4. Total the three sections from item 3 and record these in the five spaces designated "Grand Total" at the bottom left of the form.

Step 5
Protect the Monitoring Data Form from the elements and wear. Place it in a plastic cover and store it safely until you need it to complete the Monitoring Summary Form. Keep monitoring sheets in a permanent file.

APPENDIX D

The Center for Holistic Resource Management

In order to increase knowledge of holistic resource management and to develop the HRM model further, the Center for HRM was established as an international nonprofit corporation in 1984. Based in Albuquerque, New Mexico, the Center has as its goal the improvement of the human environment and quality of life through better resource management. In 1989 membership was approximately two thousand, representing fourteen different countries. Branches had been formed in Texas, New Mexico, Oregon/Washington, Nebraska, Arizona, Colorado, Montana, North Dakota, Canada, Mexico, and Namibia.

The Center offers a variety of programs designed to help families, communities, and governments use the HRM model to analyze their policies and proceed toward their goals in an ecologically, socially, and financially sound way. Courses are held three to four times each month in various locations and are attended by approximately one thousand farmers, ranchers, foresters, wildlifers, environmentalists, and government and university extension agents each year. A three-year degree program is also offered to a limited number of students per year with the Center acting as a "university without walls."

The Center sponsors an international intensive training program, which works with foreign governments and international agencies to train teachers and extension advisers. Representatives from Zimbabwe, Tunisia, Morocco, Jordan, Algeria, and the Navajo Nation have participated in this program funded by the United Nations, the U.S. Agency for International Development, the Ford Foundation, or private individuals.

A quarterly newsletter links members and practitioners together and provides a forum in which they can share problems and exchange ideas. Staff members also assist practitioners in the field and work closely with state and regional branches to organize local management clubs (self-help groups) and provide local introductory seminars.

For further information, contact the Center for Holistic Resource Management, P.O. Box 7128, Albuquerque, NM 87194.

GLOSSARY

Aide Memoire for Biological Planning. As the variables involved with crops, wildlife, weather, livestock requirements, and much more are so great, no "prescription" for the grazing of livestock is ever sound. The *Aide Memoire* (French for memory aid) describes a simple but sophisticated planning process that covers all events and their ramifications in a step-by-step process toward predetermined three-part goals.

Animal-days per Acre (ADA). A term used to express simply the volume of forage taken from an area in a specified time. It can relate to one grazing in a *paddock* or to several, in that more grazings than one can be added to give a total ADA figure. The figure is arrived at by a simple calculation as follows:

$$\frac{\text{animal numbers x days of grazing}}{\text{area of land}} = \text{ADA}$$

Animal Impact. The sum total of the direct physical influences herding animals have on the land—trampling, dunging, urinating, salivating, rubbing, digging, etcetera.

Brittle and Nonbrittle Environments. All environments, regardless of total rainfall, fall somewhere on a continuous scale from brittle to nonbrittle. Completely *nonbrittle environments* are mainly characterized by: (1) reliable precipitation regardless of volume; (2) good atmospheric moisture distribution through the year as a whole; (3) a high rate of biological decay in old plant and animal material, which is fastest near the soil surface; (4) speedy successional community development from smooth and sloped soil surfaces; (5) the development of complex and relatively stable communities with a lack of disturbance for many years.

Completely *brittle environments,* on the other hand, are characterized by: (1) unreliable precipitation, regardless of volume; (2) poor distribution of atmospheric moisture through the year as a whole; (3) in the absence of large herbivores and their predators: a) a high rate of chemical (oxidation) and physical (weathering) decay in old plant and animal material that is generally slow and moves from upper parts downward; b) a lack of disturbance, very slow successional development from bare and smooth soil surfaces, often stopping at algal capping and on steep slopes not even reaching

algal stages; and c) simpler, less diversified, and less stable successional communities.

Capping, Immature. A soil surface that has sealed with the last rainfall and on which there is no visible sign of successional movement. Capping is initiated by raindrop impact on bare soil that causes soil crumb structure to be lost.

Capping, Mature. An exposed soil surface on which *succession* has proceeded to the level of an algae-, lichen-, or moss-dominated community and stalled there.

Cell. See *Grazing Cell.*

Closed Plan. The plan made for grazing livestock through the dormant or dry season when plants do not regrow between grazings. Grazings are portioned out to the theoretical point at which no forage would be left to graze. However, that theoretical point is placed well after the time new growth is actually expected the following growing season. The plan is made toward the end of the growing season when forage reserves available during the nongrowth period become known.

Desertification. A process characterized by complete shifts in plant, animal, and soil communities. Symptoms include increased incidence of flood and drought, declining levels of soil organic matter, increased soil surface exposure, and erosion.

Energy Flow. The flow of energy from the sun to green, growing plants, then to the animals that eat the plants, and finally to the microorganisms that feed on the decaying plants and animals. Sunlight energy is indestructible, but it does continually change in form as it passes from one organism to another. As it changes form, part of it is converted to heat that is unavailable to living organisms. The volume of energy flow on the land is vital to farmers and ranchers, in particular, as it is this, rather than crops or livestock, that earns them their living.

Follow-through Grazing. A grazing strategy in which one herd of livestock grazes a paddock and is followed immediately by another herd. It is normally used to enhance the plane of nutrition of the lead herd, while the following herd is kept moving as rapidly as available paddock numbers allow. Most commonly practiced with two herds, it can also be practiced with three.

Grazing Cell. An area of land planned for grazing management purposes, normally as one unit to ensure adequate timing of grazing and recovery periods. In its most common form a grazing cell is divided into smaller units of land (*paddocks*) by radiating fences (or markers herders can use) from a central point. A grazing cell can, however, utilize any design of fencing and shape of paddocks. A grazing cell will normally contain stock year-round or at least for prolonged periods of time,

Grazing, Frequent. Grazing in the growing season that takes place with short intervals between actual grazings on the plant. With most plants frequent grazing is not harmful as long as the degree of defoliation is light.

Grazing, Severe. Grazing that removes a high proportion of a plant's leaf in either the growing or the nongrowing season. In the growing season this causes a temporary growth setback to the plant. In *brittle environments* severe grazing at some stage in the year is often essential for grass health.

Herd Effect. The impact on soil and vegetation produced by a large herd of animals in an excited state. Herd effect is not to be confused with *stock density,* as they are different although often linked. You can have high herd effect with very low stock density (such as the bison of old that ran in large herds at very low stock density, as the whole of North America was the paddock). You can have very high stock density with no herd effect, as when two steers are placed in a 1-acre paddock. Herd effect is generally produced by concentration with excitement such as supplements or other attractants applied to areas of the range where required. It does not follow automatically that a cell with a large herd and high stock density will receive good herd effect.

   Note: Herd effect is due to animal behavior and usually has to be brought about by some management action. It is mainly used as a means of applying high *animal impact.* Applied too long, on many soils it will result in surface powdering, which is undesirable.

Low-density Grazing. (Sometimes called *patch* or *selective grazing.*) This refers to the grazing of certain areas while others are left ungrazed to become stale and moribund. Normally it is caused by stock grazing at too low a *stock density,* too small a herd, or a combination of these with too short a time in the *paddock.* Once it

has started, it tends to get progressively worse, even after only one grazing, as the nutritional contrast between grazed and ungrazed areas increases with time. It is commonly corrected by forcing the stock to graze nonselectively. This is wrong and causes stock stress. Low-density grazing is rectified by increasing stock density rather than grazing pressure in the long run. In the short term it can be corrected through a number of palliatives, such as *herd effect,* fire, and grazing planning. Bear in mind that the productive and more stable ranges of the past had no problem with selective grazing even though they evolved with herding animals, which were as selective in their diets as their domestic relatives.

Mineral Cycle. The cycling of minerals fom soil to aboveground plants and animals and back to the soil again. A healthy and productive environment will promote the movement of minerals from deep soil layers to aboveground plants with a minimum of mineral loss from soil erosion or mineral leaching.

Open-ended Plan. The plan made for grazing livestock through the growing season when plants regrow between grazings. Livestock moves are planned well ahead (3 to 4 months) but remain open-ended as one cannot predict conditions precisely. Normally as the growing season continues one can anticipate more rather than less forage. However, if drought develops, that requires immediate replanning. The initial plan is made at least one month prior to the onset of the growing season.

Overgrazing. Grazing during active growth that is both severe and frequent. Generally, this results in eventual death of the plant. In intermediate stages, it results in reduced production. Overgrazing damages plants to varying degrees by utilizing energy temporarily obtained by the plant from roots sacrificed for that purpose.

Overrest. Rest of any perennial plant that is so prolonged that accumulating old material hampers growth or kills the plant. It occurs mainly in *brittle environments* where the decay process is slow.

Paddock. A smaller division of land within a *grazing cell* in which stock are grazed for short periods of time. (The American term *pasture* is deliberately not used, as in most countries it means a planted grass sward.) Paddocks can be fenced permanently, temporarily, or marked in various ways for herding without fencing.

Rest. Prolonged nondisturbance to soils and plant/animal communities. A lack of physical disturbance and fire. Partial rest takes place when grazing and browsing herding animals are on the land, but without a full complement of predators to excite them. It commonly results in damaged algae/lichen communities but no successional advance beyond to more complex, stable communities.

Stock Density. The number of animals run on a small unit of land *(paddock)* at a given moment of time. This could be from one day or less to several days. Usually expressed as the number of animals (of any size or age) run on one acre.

Stocking Rate. The number of animals run on a unit of land expressed usually in the number of acres required to run one full-grown animal throughout the year or part thereof.

Succession. The process of change and development in communities of living organisms. *Low successional:* Simple communities composed of populations of only a few species. Usually highly unstable and vulnerable. Prone to serious upheavals and fluctuations in numbers. *High successional:* Complex communities composed of populations of a great many different species of plants, animals, birds, insects, and microorganisms. Usually highly stable and not prone to high fluctuations in numbers of individual populations.

Transect. An area of land on which random measurements are made to monitor any changes arising from management practices.

Water Cycle. The movement of water from the atmosphere to the soil (or the oceans) and eventually back to the atmosphere again. An effective water cycle is one in which plants make maximum use of rainfall. Little evaporates directly off the soil, and any runoff causes no erosion and remains clear. Also a good air-to-water balance exists in the soil, enabling plant roots to absorb water readily.

INDEX

ADA. *See* Animal-days per acre
Advertising, 7-9
Aeration, soil, 93-94, 96, 100, 117
Africa
  couch grass in, 110
  wire fences of, 153
Age, monitoring plant, 91, 112, 117, 118
Aide memoire, for biological planning, 67, 68-79, 83, 164, 172
Alfalfa, financial planning for, 14, 18, 29
Algae, 100
"All-or-nothing" grazing, 102-103
Amalgamation, herd, 56, 57, 60-61, 82, 119
Analyzing financial plan, 32
Analyzing monitoring data, 119
Angora goats, 156
Animal cycles, and land cycles, 63-66
Animal-day (AD), 40-41
Animal-days per acre (ADA), 41-42
  defined, 41, 172
  and forage quality, 72-74
  and forage reserves, 51
  on grazing record, 78, 79
  and growing season grazing, 76
  and multiple herds, 63, 64
  paddock numbers and, 45, 48, 49
  simplified grazing guidelines and, 82
  stock density and, 48, 49
  and stocking rate, 54, 76-77
Animal impact
  biological planning and, 40, 61, 71, 82
  defined, 172
  and landscape goals, 71, 103
  and mineral cycle, 96
  monitoring, 87, 96, 100, 103, 111, 118, 119
  and soil capping, 100
  transect procedure and, 111
  and water cycle, 118, 119
  See also Herd effect; Stock density
Animals
  deficiency symptoms, 96
  identifying, 100-102
  and mineral cycle, 95-96
  and succession, 91
  and water cycle, 94
  weighing, 106
  See also Animal impact;

  Livestock; Wildlife
Animal unit months (AUM), 40
Annual income and expense plan, 22, 25, 28, 31, 160
Annual plants, 91, 111
Ants, 91, 95
Arizona, dormant season, 55
Art, 39

Barlite ranch, 61
Basal cover, 109, 116, 117
Bell signal, 156
Bermuda grass, 110
Biological monitoring, 87-120
  creating plan, 106-119
  equipment for, 107
  Field Instruction Sheet, 166, 170
  forms, 107, 108-109, 110, 112, 113-119, 166-169
  gathering data, 106-115
  mastering basics, 89-105
  summarizing data, 115-119
Biological Plan and Control chart, 69-71, 76, 78, 164-165
Biological planning, 39-83, 104, 123, 136
  aide memoire for, 67, 68-79, 83, 164, 172
  chart, 69-71, 76 78, 164-165
  creating plan, 67-83
  and land planning, 39, 141, 142-144, 145
  mastering basics, 40-66
  object of, 83
  operating, 78
  simplified grazing guidelines, 67, 80-83
  warnings, 67
"Biological year" worksheet, 17-18, 19
Birds, 66, 105, 152
Bison, 64-66
Blue gramma, 98, 110
Borrowing, 7, 26
Brainstorming
  for biological planning, 68
  for financial planning, 15-16, 24
  for land planning, 137
Brittle environment
  capping in, 110
  defined, 172
  grazing patterns in, 103
  herd effect in, 61, 100
  overrest in, 99

paddock numbers in, 47
planning in, 47, 61
transect procedure in, 107, 111
Browse lines, 54, 99
Browsing, 54
  See also Overbrowsing
Brush
  herd effect and, 61, 62
  monitoring, 110
Bryce, Ebenezer, 16
Bryce Canyon, 16
Bulls, 41, 42, 151
Bunchgrasses, 98, 109

Cake, and herd effect, 62
Calves, 152
Calving, 60, 63-64, 66
Canopy, monitoring, 110, 116, 117, 118
Canyons, 146
Capital investments, 9, 139, 144
Capping, 100, 110, 117, 118
  broken, 100
  immature, 100, 172
  mature, 100, 116, 172
  recent, 100
Cargill, 9
Cash, managing, 30-32, 145
Cattle
  biological monitoring of, 87
  bulls, 41, 42, 151
  and cell centers, 150, 151
  cost of herding option with, 144-145
  dairy, 10-11, 91, 92
  destocking, 56-57
  fencing for, 153
  gross margin analysis for, 10-13
  herd size of, 132-135
  Livestock Production Plan for, 19-21
  measuring range use of, 41, 42
  standard animal units for, 41, 43
  steers, 14, 41, 42, 64, 145
  and succession, 91, 92
  yearlings, 54, 55, 64, 145
  See also Cows
Cattle-day (CD), 41
Cause and Effect guideline, 4, 5
Cells, 39, 45
  in *aide memorie*, 68-69, 72
  building, 139-145, 146-153
  centers, 136, 147-153

Cells (continued)
  corridors, 150, 151
  defined, 39, 173
  forage reserves in, 50-51
  handling livestock in, 72
  herd amalgamation in, 60
  and multiple herds, 62-63
  paddock numbers in, 45-49
  simplified grazing guidelines
    and, 81-82, 83
  size, 132-135
  wagon-wheel, 39, 132, 134, 136
Center for Holistic Resource
  Management, 171
  land planning problem designed
    at, 142
  publications/forms available from,
    16, 159, 164
  transect procedure developed at,
    106
Centers, cell, 136, 147-153
Chains, 5
  food, 96, 97
  solar, 5-7
Chart, biological planning, 69-71,
  76, 78, 164-165
Checklist
  aide memoire vs., 67
  land planning, 125-126, 139
Chemical fertilizer, 96
Cicadas, 116
Circles
  planning, 132, 133
  See also Wagon-wheel cells
Clear cutting timber, 7
Climate
  and cool- and warm-season (C/W)
    plants, 111
  and fence grounding, 154
  See also Precipitation
"Climax" communities, 90
Closed plan, 68, 172
Clover seed, 96
Common sense, 67
Communication, among land
  planners, 125
Compaction, soil, 96
Compromises, land planning and,
  125
Computer programs
  for biological planning, 74
  for financial planning, 22, 24, 32
Conception, livestock, 64, 87
Construction, 139-144, 145
Consumptive use, 7
Control, 87
  financial planning and, 16, 22, 30,
    32-34, 40
Control Sheets, 22, 34, 163
Cool- and warm-season (C/W)
  plants, 57, 111-112, 117, 118

Corridors, cell, 150, 151
Cost accounting procedures, 21-22
Costs
  financial planning and, 17, 24, 25-
    30, 32-33
  fixed, 8-11, 15
  land planning and, 139-145, 146
  production, 26
  variable, 8-11
Cotton, 15
Couch grass, 110
Cover. See Ground cover
Cows
  calving, 60, 63-64, 66
  cattle-days for, 40, 41
  dairy, 10-11, 91, 92
  destocking, 56-57
  gross margin analysis for, 10-13
  habits and routines of, 103
  herd amalgamation of, 60
  Livestock Production Plan for, 19-
    21
  multiple herds and, 62
  standard animal units and, 42
  worksheet on "biological year" of,
    17-18, 19
Coyotes, 91, 105
Crop farming
  in aide memoire, 69
  information gathering re, 127
  and mineral deficiencies, 96
  monitoring, 89
  succession in, 92
  weak link in, 6
Cyclical use, 7

Dairy cattle, 10-11, 91, 92
Dairy operations, 10-11
Dams, fencing and, 153
Dart, sampling, 109, 111
Debt, 8-9, 26, 145
Decomposers, 96, 117
Deer, 153
Deficiency symptoms, plant and
  animal, 96
Density
  plant, 106, 110, 111, 116
  stock, see Stock density
Depreciation, 30
Deseret Ranch, Utah, 64-66
Desertification, 172
Destocking, 56-57
Diesel fuel, 7
Distance, monitoring, 111, 116, 117,
  118, 119
Diversity, 112-113
  and forage planning, 54
  gross margin analysis and, 15
  mineral cycle and, 118
  and soil organisms, 95
  and succession, 15, 91, 95, 116, 117

Dogs
  for driving livestock, 62
  prairie, 91
Don Coyote (Hyde), 105
Dormant season, 55-60, 68
  animal-days and, 41
  grazing plans for, 77-78
  overstocking and, 54
  recording grazing for, 79
  simplified grazing guidelines and,
    80, 82
  stocking rate for, 76-77
Dot tally method, 109, 111
Driving livestock, 62, 156, 157
Drought, 55
  forage reserves for, 50, 68, 72, 77
  habits and routines and, 103
  herd amalgamation and, 60-61
  land planning and, 144
Dunes, and erosion, 93
Dung
  cell center and corridor and, 150,
    151
  mineral cycle and, 95, 96
  pests in, 66
  stock density and, 49

Ecosystem blocks. See Energy flow;
  Mineral cycle; Succession; Water
  cycle
Educating. See Training, livestock
Electric fencing, 153-155
Elk, 64
Emergency decisions, 17
Emergency reserves, 41, 55
  See also Drought
Employees
  NMI and, 16-17
  See also Labor
Energy flow, 87, 90, 95, 96-97, 118
  capping and, 110, 118
  defined, 96-97, 172
  mesophytic (broad-leafed) plants
    and, 95, 118
  monitoring, 87, 90, 96-97, 118
  weak link in, 5, 6, 7, 27, 140
Energy/Wealth Source and Use
  guideline, 6, 6-15, 140
  See also Solar energy
Erosion, monitoring, 92-93, 100, 113,
  116, 117, 118, 119
"Estimated profit and loss
  statement," 32
Expenses
  forms, 22, 25, 28, 31, 160
  inescapable, 27
  maintenance, 27
  planning, 17, 24, 25-30, 32-33
  wealth-generating, 27
  See also Costs

Fences, 45, 136, 139-144, 145, 146, 147
  cell center and, 148-149, 151
  electric, 153-155
  maps and, 126
  patterns, 39, 148-149
  posts, 153, 154
  and stock density, 119
  and trailing, 104, 147
  training livestock for, 155-157
  wildlife and, 54, 153
  wire, 153-155
Fertilizer, 9, 95, 96
Field Instruction Sheet, for biological
  monitoring, 166, 170
Financial planning, 3-35, 40, 123
  analyzing, 32
  creating plan, 21-34
  expenses, 17, 24, 25-30, 32-33
  forms, 19-21, 22, 24, 25, 159-163
  income, 23-24, 26, 32-33
  and land planning, 139-145, 146
  mastering basics, 4-21
  planning, 21-23
  scheduling, 23
  team, 22-23, 32, 34
  weak link-strengthening, 6, 27-28
  See also Costs; Investments
Firebreaks, herd effect and, 62
Fires, 66, 89, 91
Fir trees, 91
Fixed costs, 8-11
  defined, 9, 10
  overhead, 8-11, 15
Flagging, paddock designation by,
  146
Flexibility guideline, 123
Floods, 66, 89
Flow patterns
  erosion and, 92-93, 113
  See also Energy flow
Follow-through grazing, 62-63, 72,
  83, 172
Food chain, and energy flow, 96, 97
Forage
  canopy and, 118
  and cool- and warm-season (C/W)
    plants, 111
  cycles, 64, 65
  during dormant season, 55-60
  during drought, 60-61
  measuring, 40-54
  planning, 40-61, 64, 65, 68, 72-74
  quality ratings, 72-74
  reserves, 41, 50-54, 55, 60, 68,
    72, 77, 80
  See also Grazing
Forbs, 110
  cool-season, 57
  overbrowsing and, 98-99
  and succession, 92
Forest, 7, 89, 95

Form, plant, 98-99, 112, 117, 118
Forms
  biological monitoring, 107, 108-109,
    110, 112, 113-119, 166-169
  biological planning, 164-165
  Control Sheets, 22, 34, 163
  financial planning, 19-21, 22, 24,
    25, 159-163
  Livestock Production Plan, 19-21,
    22, 24, 162
  See also Worksheets

Gardzia, Kirk, 106
Gasoline use and purchase,
  worksheets on, 19, 29
Gates, 155
Goals, 4, 23-24
  See also Landscape; Production;
    Quality of life
Goats, 62, 64, 91, 92, 156
Gong signal, 157
Grasses
  cool-season, 57
  herd effect and, 61, 62
  monitoring, 91, 95, 98, 100, 109,
    110, 111
  sod-bound, 100, 110, 111
  See also Forage
Grasshoppers, 105
Grazing, 40-83
  "all-or-nothing," 102-103
  and animal cycles, 63-66
  and canopy, 110
  and cool- and warm-season (C/W)
    plants, 111
  during dormant season, 55-60
  during drought, 60-61
  follow-through, 62-63, 72, 83, 172
  frequent, 173
  in growing season, 75-76, 78
  land planning and, 136, 150
  low-density, 102-103, 173
  measuring forage for, 40-54
  monitoring of, 87, 102-103
  multiple herd, 62-63
  in nongrowing season, 77-78
  "patch," 102-103, 173
  patterns, 102-104
  selective, 173
  severe, 49, 173
  simplified guidelines, 67, 80-83
  See also Grazing periods;
    Overgrazing
Grazing cells. See Cells
Grazing periods, 45
  fence planning and, 141, 142-144
  herd amalgamation and, 60
  maximum, 74-75, 78, 80, 81
  minimum, 74-75, 80, 81
  for multiple herds, 62-63, 64, 75
  in nongrowing season, 78

and overgrazing, 45, 98, 141
  paddock numbers and, 47, 49, 136,
    141
  simplified grazing guidelines and,
    81
  stocking rate and, 53, 54
Greasewood, 92
Gross margin analysis, 4, 9-15, 24
  danger of, 15
Ground cover, 109
  monitoring, 92, 109, 116, 117, 118
Grounding, with electric fencing,
  154-155
Group sessions
  land planning, 125-126, 136-139
  See also Team planning
Growing season
  biological planning during, 68
  grazing plans for, 75-76, 78
  overstocking in, 54
  recording grazing for, 79
  simplified grazing guidelines and,
    80, 81
  stock density during, 57
Growth rates, 44, 47
  monitoring, 87, 106
  recording, 79
  See also Recovery periods
Guest, Charles, 61
Guest, Katie, 61
Guidelines, 4-15
  for simplified grazing, 67, 80-83
Gullies
  fence crossings, 155
  herd effect and, 62

Habitat
  dry, 95
  middle, 95
  monitoring, 95, 110, 116, 117, 118
  wet, 95
Habits, livestock, 103-104
Hardware
  in land planning, 146-157
  See also Fences
Hawaii, 132
Hay
  financial planning for, 29
  and herd effect, 62
  training with, 157
  winter, 66, 103
Herbivores, 96
Herd effect, 40, 61-62, 119
  and capping, 10
  defined, 61, 173
  signals and, 156
Herds
  amalgamation of, 56, 57, 60-61, 82,
    119
  land planning and, 127, 132-135,
    144-145, 146

Herds *(continued)*
  multiple, 62-63, 64, 75, 83
  simplified grazing guidelines and,
    82, 83
  size, 132-135
Historical data, 89-90
Holism, 3, 102, 104
  and fencing, 123
  and monitoring, 87
Holistic Resource Management
  (Savory), 4
Horn signal, 156-157
Horses, 92
  semiferal, 57
Human creativity, 24
Hunting, 54
Hyde, Dayton, 105
Hydrophytic plants, 95

Identifying species, 100-102, 112-113,
  117, 118
Income
  forms, 22, 25, 28, 31, 33, 160
  halving, 26
  net managerial, 16-17
  planning, 23-24, 26, 32-33
Infestation, 62, 116
Infiltration, soil, 93-94
Information gathering
  biological monitoring, 106-115
  land planning, 125-126, 127
Investments
  capital, 9, 139, 144
  reinvestment, 5
Irrigated zones, 136

John Deere, 9

Kroos, Roland, 142

Labor
  financial planning and, 16-17
  and follow-through grazing, 62
  land planning and, 144
Lactation, livestock, 64
Lambing, 60, 63
Land
  boundaries, 132-133
  as fundamental resource, 23
  monitoring, 87, 106
  See also Forage
Land cycles, and animal cycles, 63-66
Land division, 45-49, 146
  See also Fences
Land planning, 39, 123-157
  creating plan, 136-157
  information collection, 125-126, 127
  maps, 126-132, 136, 137-138
  mastering basics, 125-135
Landscape, 4, 40
  biological monitoring and, 87-88,

90, 103, 110, 116
  biological planning and, 45, 68, 71,
    80
  financial planning and, 24
Layouts, in land planning, 136, 146-
  157
Leaching, 95, 96
Legumes, 96
Lichen, 100
Liquid mineral licks, and herd effect,
  62
Litter
  banks, 93
  breakdown of, 95
  as ground cover, 109, 116, 117
  herd effect and, 61
  and mineral cycle, 95
  and soil capping, 100
  stock density and, 102
  and water cycle, 93, 118
Liver flukes, 66
Livestock
  biological planning and, 39-83
  concentrating, 136 (*see also*
    Amalgamation, herd)
  conception, 64-87
  and cool- and warm-season (C/W)
    plants, 111
  driving, 62, 156, 157
  financial planning and, 5-6, 14, 19-
    21, 22, 24
  habits and routines of, 103-104
  land planning and, 136, 147, 151-
    152, 155-157
  monitoring, 87, 95-96, 98, 102-104
  success and, 92
  See also Cattle; Goats; Sheep
Livestock Production Plan, 19-21, 22,
  24, 162
Living organisms. *See* Organisms
Los Ojos Ranch, 54-55
Low-density grazing, 102-103, 173

Machinery, 9, 30
Management concerns, in Biological
  Plan and Control Chart, 69-70
"Management control," 16
Management needs (special), in
  Biological Plan and Control
  Chart, 71
Managers, NMI and, 16-17
Managing cash, 30-32, 145
Manure. *See* Dung
Manure spreader, 96
Maps, 126-132
  blank, 137-138
  grid for, 129
  overlays, 126-132, 139
  sketched, 136, 138
  topographic, 129-130, 136, 138
Marfa, Texas, Barlite ranch, 61

Marginal Reaction guideline
  financial planning and, 7-9, 27, 28
  land planning and, 132, 139, 140,
    142, 144
Marketing
  gross margin analysis and, 15
  of solar energy, 5, 106
  as weak link, 5, 6, 7-9
  wealth generation and, 25
Matthews, Ivey, 103
Measuring
  map acreage, 128-132
  range use and forage, 40-54
Mental attitude, 22
Mesophytic, 95
Mesophytic plants, 95
Mineral cycle, 87, 90, 94-96
  capping and, 100, 110, 117
  defined, 173
  and succession, 91, 95, 96
  summarizing data re, 117-118
Mineral supplements
  and herd effect, 62
  monitoring, 95-96
Mining, 58, 87
Molasses
  fence training with, 155-156
  and herd effect, 62
Monitoring. *See* Biological
  monitoring
Monitoring Data Form, 166
  blank, 167-168
  completed, 114
  instructions, 107, 108-109, 110, 112,
    113-115
Monitoring Summary Form, 115-119,
  166, 169
Moose, 91
Mormon tea, 92
Mortgage payments, 11
Moss, 100
Multiple herds, 62-63, 64, 75, 83

Natural barriers, for land division,
  45, 146
Navajos, 101-102, 126
Net managerial income (NMI), 16-17
Neutral ground, land planning and,
  125
New Mexico
  Los Ojos Ranch, 54-55
  transect markers on range in, 107
New Zealand, energizers from, 153
Nitrogen, 96
NMI (net managerial income), 16-17
Nonbrittle environment
  defined, 172
  grazing patterns in, 102, 103
Nongrowing seasons
  grazing plans for, 77-78

Nongrowing seasons *(continued)*
  simplified grazing guidelines and, 80, 82
  stocking rate for, 76-77
  See also Dormant season
North Carolina, dairy operations, 10-11
Nutrition, 64, 65
  in dormant season, 57-60
  in drought, 61
  herd amalgamation and, 61
  for multiple herds, 62
  and paddock numbers, 82
  See also Forage; Supplements

Open-ended plan, 68, 173
Optimism trap, 26
Organisms
  monitoring, 104-105, 110, 116, 117, 118
  soil, 95, 96, 117
Overbrowsing, plant form with, 98-99, 112, 117, 118
Overdraft arrangements, 31, 32
Overgrazing, 39, 44, 45, 55
  and cool- and warm-season (C/W) plants, 57, 111
  defined, 173
  fencing and, 140-141
  paddock numbers and, 47, 49, 78
  plant form with, 98, 112, 117, 118
  routines and, 103
  and simplified grazing guidelines, 80, 81
  stock density and, 49
Overhead, 8-11, 15
Overrest
  defined, 173
  plant form with, 98, 99, 112, 117, 118
  and productivity, 50
Overstocking
  biological planning and, 53, 54, 55
  grazing patterns with, 102-103
  simplified grazing guidelines and, 80, 82
"Owner control," 16

Paddocks
  ADA measurement in, 41-42
  in *aide memoire*, 68-79
  in Biological Plan and Control Chart, 70
  building, 139-145, 146-147, 153, 156
  cell centers and, 147, 151
  and cell size, 132
  defined, 173-174
  in dormant season, 56-60
  fencing, 147, 153, 156
  grazing days recorded in, 81
  grazing patterns in, 102, 103

in growing season, 75-76, 78, 79
habits and routines with, 103-104
herd amalgamation in, 60
layouts, 136, 146-147
and mineral cycle, 96
and multiple herds, 62-63, 64
in nongrowing season, 76-77, 78
numbers, 44, 45-49, 59, 78, 103, 136, 141
and overgrazing, 44
productivity ratings, 72-74
reassessing, 82
sequence grazed, 66
simplified grazing guidelines and, 81-82
and stock density, 48-49
stocking rate in, 53, 54
temporary fencing, 147
and timing, 45-48
Parasites, 6
Pasture
  financial planning and, 14
  overrest and, 99
"Patch" grazing, 102-103, 173
Pedestaling, of plants and rocks, 92
Perennial plants, 91, 92, 98, 110, 111, 113, 119
Permeability, soil, 93-94
Pests, 66, 105
Photographs, monitoring with, 89, 90, 108, 110, 119
Planimeter, 132
Planning, 45, 87
  See also Biological planning; Financial planning; Land planning
Plants, 91-92
  age of, 91, 112, 117, 118
  annual, 91, 111
  cool- and warm season (C/W), 57, 111-112, 117, 118
  dead centers, 98, 99
  deficiency symptoms, 96
  density, 106, 110, 111, 116
  distorted, 98, 99
  and energy flow, 95, 96, 118
  form, 98-99, 112, 117, 118
  growth rates, 44, 47, 79, 87, 106
  hydrophytic, 95
  identifying, 100-102, 112-113, 117, 118
  and mineral cycle, 95, 96
  overbrowsed, 98-99, 112, 117, 118
  overgrazed, *see* Overgrazing
  overrested, *see* Overrest
  pedestaling, 92
  perennial, 91, 92, 98, 110, 111, 113, 119
  poisonous, 57, 66, 76
  root depth, 95, 96, 117, 118
  root systems, 99
  summarizing data re, 116-118

transect procedure and, 106, 109, 110, 111
type, 110, 116, 117, 118, 119
and water cycle, 94, 95
woody, 92, 99 (*see also* Trees)
xerophytic, 95
See also Forage
Plumb bob, 109
Points
  hits, 109, 111
  sampling, 106-107, 109, 111
  water, 126, 140, 150
Poisonous plants, 57, 66, 76
Poisons, for pests, 105
Pond, 152
Population management. *See* Stocking rate
Posts, fence, 153, 154
Prairie dogs, 91
Precipitation
  capping and, 100
  effective, 72, 93-94
  recording, 79
  and recovery period, 72
  See also Rain; Snow
Predators, 105
  biological planning for, 66
  energy flow and, 96
  and herd effect, 62
  succession and, 116
Prickly pear, 95
Product conversion link, 5, 6, 7
Production
  biological monitoring and, 103, 116
  biological planning and, 45, 68, 79
  costs of, 26
  financial planning and, 4, 8-9, 23-24, 26
  scale of, 8-9
  See also Profit
Productivity
  and fencing, 142
  overrest and, 50
  paddock, 72-74
Profit
  as goal, 4, 23-24
  gross margin analysis and, 10-11, 15
  optimism trap re, 26
Pronghorn, 54

Quality, forage, 72-74
Quality of life, 4, 9, 15, 116

Raccoons, 105
Rain
  canopy and, 110, 116, 117
  and cool- and warm-season (C/W) plants, 111
  effective, 93-94
  and grazing patterns, 102, 103

Randomness, of point samples, 106-107
Rangelands, monitoring, 89
Record keeping
    grazing, 78-79
    photographic, 108
Recovery periods, 45
    controlling, 136
    deciding range of, 72, 80
    herd amalgamation and, 60
    land planning and, 136, 141
    and maximum/minimum grazing periods, 74
    multiple herds and, 62, 63
    paddock numbers and, 45, 46, 47, 49, 103
    recording, 79
    simplified grazing guidelines and, 80
Redirecting policy, 119
Reinvestment, 5
Rest
    biological planning and, 40
    defined, 174
    See also Overrest
Riparian zones, 136, 139, 142, 144
Risk, 8-9
Rituals, 15
Roads, on map overlays, 126
Roadsides, herd effect and, 62
Rocks
    and energy flow, 118
    as ground cover, 109, 118
    paddock designation by, 146
    pedestaling, 92
Rodents, 91, 95, 105
Root depth, varying, 95, 96, 117, 118
Root systems, weakened, 99
Rotation
    during dormant periods, 58-60
    routine, 104
    thoughtless, 78
    See also Grazing periods
Routines, livestock, 103-104
Rye, 66

Sackville-West, V., 21
Salt, and herd effect, 62
Salt grass, 95
Sampling, point, 106-107, 109, 111
Savory, Allan, 4, 16, 17, 21, 35, 39
Scale, of production, 8-9
Scheduling
    biological monitoring, 108
    financial planning, 23
    land planning, 145-146
    See also Time
Season
    biological monitoring, 108
    biological planning, 68

See also Dormant season; Growing season
Seasonal diversity, 91
Sedges, herd effect and, 62
Seeps, water, 152
Sheep
    and fences, 153
    gross margin analysis and, 11-13
    lambing, 60, 63
    nutritional needs of, 64
    sheep-days for, 40, 41
    standard animal units for, 42
    stocking rate for, 52
    succession and, 91, 92
Sheep-day (ShD), 40, 41
Shortcuts, monitoring, 119
Shrubs, overbrowsing and, 98-99
Signal, animal training, 156-157
Siltation, and erosion, 93
Simplified grazing, guidelines for, 67, 80-83
Siting, cell center, 150
Size, herd/cell, 132-135
Sloughs, fencing and, 153
Snow
    and cool- and warm-season (C/W) plants, 111
    grazing in, 54, 64-66
Society and Culture guideline, 4, 5, 124
Sod-bound condition, 100, 109, 110, 111, 116
Soil
    acidity, 96
    aeration, 93-94, 96, 100, 117
    alkalinity, 96
    capping, 100, 110, 116, 117, 118, 172
    compaction, 96
    covered, 110, 116
    and energy flow, 96
    erosion, 92-93, 100, 113, 116, 117, 118, 119
    herd effect and, 62
    infiltration, 93-94
    and mineral cycle, 95, 96
    organisms, 95, 96, 117
    permeability, 93-94
    pH, 96
    shortcuts and, 119
    sodium, 96
    transect procedure and, 106
Solar energy, 5-7, 27, 40
    marketing of, 106
    wealth generation with, 6-7, 27, 32
    See also Energy flow
Southwest, American
    in dormant season, 55, 57
    pedestaling in, 92
    See also New Mexico
Species
    diversity, see Diversity

identifying, 100-102, 112-113, 117, 118
Splash patterns, and erosion, 93
Spreadsheet programs, 22, 24
Springs, monitoring, 89
Squirrels, 91, 105
Stagnation, in brittle environment, 99
Standard animal units (SAU), 41, 42-43, 76
Steers, 14, 41, 42, 64, 145
Stock day (SD), 41
Stock density, 40, 119
    defined, 48, 51, 174
    in drought, 60-61
    and grazing patterns, 102-103
    in growing season, 57
    and multiple herds, 64
    paddock layout and, 147
    paddock numbers and, 47-49
    simplified grazing guidelines and, 82
    and stocking rate, 51, 54, 119
    See also Overstocking
Stocking rate, 40, 51-55, 69
    animal-days and, 41
    defined, 174
    figuring, 51-52, 76-77
    land planning and, 144-145
    simplified grazing guidelines and, 80, 82
    stock density and, 51, 54, 119
Streams
    fence crossing, 155, 156
    monitoring, 89
    for water supply, 151
Stress, animal, 78, 103
Subchains, 5
Succession, 66, 87, 90-92
    capping and, 100, 110, 116
    cool- and warm-season (C/W) plants and, 111, 117
    defined, 174
    diversity and, 15, 91, 95, 116, 117
    grazing patterns and, 102
    herd effect and, 61, 62
    and living organisms, 104, 116
    and mineral cycle, 91, 95, 96
    overgrazing and, 98, 117
    overrest and, 99, 117
    summarizing data re, 116-117
    transect procedure and, 106, 111, 112
    and water cycle, 91, 94
Sunlight energy chain, 5-6
    See also Solar energy
Supplements
    and fence training, 157
    and herd amalgamation, 61
    and herd effect, 62
    monitoring, 95-96

Tank, water, 152
Taxes, 11
Team planning
  financial, 22-23, 32, 34
  land, 137-139
Texas, Barlite ranch, 61
Timber, clear cutting, 7
Time, 45
  for biological monitoring, 108
  biological planning and, 40, 45-49,
    55, 60, 68, 80
  financial planning and, 9, 15, 17, 23
  land planning and, 125, 145-146
  See also Grazing periods; Recovery
    periods; Season
"Time reserve"
  biological planning for, 72, 77, 80
  See also Drought
Topographic maps, 129-130, 136, 138
Trailing, 104, 147, 150
Training, livestock
  fence, 155-157
  and herd effect, 62, 119, 156
Transects, 89-90, 106-119, 174
Trees
  as fence posts, 153
  monitoring, 91, 92, 98-99, 110
  overbrowsing and, 98-99
  and succession, 91, 92
Troughs, water, 152

U.S. Bureau of Land Management, 13
U.S. Geological Survey "Seven
  Minute Series" map, 128

U.S. Soil Conservation Service, 112
Utah, Deseret Ranch, 64-66

Variable costs, 8-11
  defined, 10
Visual demarcation, for land
  division, 45

Wagon-wheel cells, 39, 132, 134, 136
Wallace, David, 10
Water cycle, 87, 90, 92-94, 95
  and animal impact, 118, 119
  capping and, 100, 110, 117
  defined, 174
  erosion and, 116
  and succession, 91, 94
  summarizing data re, 117
Water points, land planning and,
  126, 140, 150
Watersheds
  land planning and, 139
  monitoring, 89
Water supply, 139
  in cell, 151-153
  and cell size, 132
Weak link, 4, 5-7, 27-28, 139
  energy conversion as, 5, 6, 7, 27, 140
Wealth, 23
  generation of, 6-7, 21-22, 25-26, 27,
    139
  solar, 6-7, 27, 32
  See also Energy/Wealth Source and
    Use guideline
Weeds, herd effect and, 62

Wetlands, 105, 117
Whips, for driving livestock, 62
Whistle signal, 156-157
Whole Ecosystem guideline, 4, 5
Wilderness, as production goal, 23
Wildlife, 105
  and fences, 54, 153
  forage for, 41, 54, 55, 57, 58, 68, 72,
    80
  and herd effect, 62
  monitoring and, 89, 105
  simplified grazing guidelines and,
    80, 82, 83
  water seep for, 152
  See also Predators
Wire fences, 153-155
Woody plants, 92, 99
  See also Trees
Worksheets, 17-19, 22, 24, 161, 164
  alfalfa production and sales, 18, 29
  "biological year," 17-18, 19
  construction project, 145
  gasoline use and purchase, 19, 29
  income and expenses, 17, 25, 26,
    29, 30, 160
  Livestock Production Plan, 20-21,
    22, 162
Worms, 95

Xerophytic plants, 95

Yamsi Ranch, 105
Yearlings, 54, 55, 64, 145
Yucca, 95

ABOUT THE AUTHORS

Allan Savory was born in Rhodesia (today Zimbabwe) in 1935, and graduated from the University of Natal (South Africa) in 1955. After seven years as a research officer in the game department, Savory left government service to farm and ranch and to establish a consulting service based on discoveries he had made connecting the health of land with wild animal populations.

From 1956 until 1972, Savory served in the Rhodesian army; additionally, he entered Parliament in 1968. In 1974, Savory became president of the National Unifying Force, an alliance of moderate white parties; he led this group until he was exiled by the Ian Smith government in 1978.

When exiled, Savory moved his international consulting service to the United States. In 1984, with a group of ranchers, environmentalists, educators, and others concerned with America's deteriorating resource base, he established the international nonprofit Center for Holistic Resource Management in Albuquerque, New Mexico. The purpose of the center is to coordinate the development of holistic management and to provide training in this new management technique.

A graduate of Yale University, Sam Bingham is a freelance writer specializing in agriculture and land-use issues. A former reporter for *Newsweek* magazine and associate editor for *The Atlantic,* Bingham has published articles in a wide variety of newspapers and periodicals. In addition, he is a regular contributor to *High Country News* and the *Denver Post.* Bingham has published three previous books: *Navajo Farming, Between Sacred Mountains: Navajo Stories and Lessons from the Land* (both coauthored with Janet Bingham), and *Living from Livestock,* a textbook for Navajo stock growers and vocational agricultural teachers.

Sam Bingham first encountered Holistic Resource Management and Allan Savory while he and his wife were teaching on the Navajo Indian Reservation in Arizona in 1980. He has been active in the Center for Holistic Resource Management and the local Colorado chapter of HRM. In 1984, Allan Savory asked Bingham to work with him on the textbook *Holistic Resource Management,* which provides the theory and origins of Savory's comprehensive planning model. The *Holistic Resource Management Workbook* is the companion volume to the textbook, which was published by Island Press in 1988.